ELECTROMAGNETICS EXPLAINED

A HANDBOOK FOR WIRELESS/RF, EMC, AND HIGH-SPEED ELECTRONICS

Other titles in the EDN Series for Design Engineers

Analog and Digital Filter Design, Second Edition, by Steve Winder
0-7506-7547-0, Paperback, 512 pgs., $59.99

Practical RF Handbook, Third Edition, by Ian Hickman
0-7506-5369-8, Paperback, 304 pgs., $39.99

Power Supply Cookbook, Second Edition, by Marty Brown
0-7506-7329-X, Paperback, 336 pgs., $39.99

Radio Frequency Transistors, Norman Dye and Helge Granberg
0-7506-7281-1, Paperback, 320 pgs., $49.99

Troubleshooting Analog Circuits, by Robert A. Pease
0-7506-9499-8, Paperback, 217 pgs., $34.99

The Art and Science of Analog Circuit Design, edited by Jim Williams
0-7506-7062-2, Paperback, 416 pgs., $34.99

Introducing the EDN BOOKSTORE

The EDN Bookstore offers you:

- Hundreds of great technical books by engineers for engineers
- Monthly drawings for FREE books
- Specials offers & discount pricing
- FREE sample chapters

Visit the EDN Bookstore at www.ednmag/bookstore.com

ELECTROMAGNETICS EXPLAINED

A HANDBOOK FOR WIRELESS/RF, EMC, AND
HIGH-SPEED ELECTRONICS

Ron Schmitt

Newnes
An imprint of Elsevier

Amsterdam   Boston   London   Oxford   New York   Paris
San Diego   San Francisco   Singapore   Sydney   Tokyo

Newnes is an imprint of Elsevier Science.

Copyright © 2002 by Elsevier Science (USA)

All rights reserved.

No part of this publication may be reproduced, stored in a retrieval system, or transmitted in any form or by any means, electronic, mechanical, photocopying, recording, or otherwise, without the prior written permission of the publisher.

Permissions may be sought directly from Elsevier's Science and Technology Rights Department in Oxford, UK. Phone: (44) 1865 843830, Fax: (44) 1865 853333, e-mail: permissions@elsevier.co.uk. You may also complete your request on-line via the Elsevier homepage: http://www.elsevier.com by selecting "Customer Support" and then "Obtaining Permissions".

 Recognizing the importance of preserving what has been written, Elsevier Science prints its books on acid-free paper whenever possible.

Library of Congress Cataloging-in-Publication Data
Schmitt, Ron.
    Electromagnetics explained: a handbook for wireless/RF, EMC, and high-speed electronics / Ron Schmitt.
      p. cm.
    Includes bibliographical references and index.
    ISBN 0-7506-7403-2 (hc.: alk. paper)
    1. Electronics. 2. Radio. 3. Electromagnetic theory. I. Title.
TK7816 .S349 2002
621.381—dc21
                                                                                 2001055860

British Library Cataloguing-in-Publication Data
A catalogue record for this book is available from the British Library.

The publisher offers special discounts on bulk orders of this book.
For information, please contact:

Manager of Special Sales
Elsevier
200 Wheeler Road
Burlington, MA 01803
Tel: 781-313-4700
Fax: 781-313-4880

For information on all Elsevier publications available, contact our World Wide Web home page at: http://www.newnespress.com

10 9 8 7 6 5 4

Printed in the United States of America

CONTENTS

PREFACE xi

ACKNOWLEDGMENTS xv

1 INTRODUCTION AND SURVEY OF THE ELECTROMAGNETIC SPECTRUM 1
The Need for Electromagnetics 1
The Electromagnetic Spectrum 3
Electrical Length 8
The Finite Speed of Light 8
Electronics 9
Analog and Digital Signals 12
RF Techniques 12
Microwave Techniques 16
Infrared and the Electronic Speed Limit 16
Visible Light and Beyond 18
Lasers and Photonics 20
Summary 21

2 FUNDAMENTALS OF ELECTRIC FIELDS 25
The Electric Force Field 25
Other Types of Fields 26
Voltage and Potential Energy 28
Charges in Metals 30
The Definition of Resistance 32
Electrons and Holes 33
Electrostatic Induction and Capacitance 34
Insulators (Dielectrics) 38
Static Electricity and Lightning 39
The Battery Revisited 45
Electric Field Examples 47
Conductivity and Permittivity of Common Materials 47

3 FUNDAMENTALS OF MAGNETIC FIELDS 51
Moving Charges: Source of All Magnetic Fields 51
Magnetic Dipoles 53
Effects of the Magnetic Field 56
The Vector Magnetic Potential and Potential Momentum 68
Magnetic Materials 69
Magnetism and Quantum Physics 73

4 ELECTRODYNAMICS 75
Changing Magnetic Fields and Lenz's Law 75
Faraday's Law 76
Inductors 76
AC Circuits, Impedance, and Reactance 78
Relays, Doorbells, and Phone Ringers 79
Moving Magnets and Electric Guitars 80
Generators and Microphones 80
The Transformer 81
Saturation and Hysteresis 82
When to Gap Your Cores 82
Ferrites: The Friends of RF, High-Speed Digital, and Microwave Engineers 83
Maxwell's Equations and the Displacement Current 84
Perpetual Motion 86
What About D and H? The Constituitive Relations 87

5 RADIATION 89
Storage Fields versus Radiation Fields 89
Electrical Length 91
The Field of a Static Charge 94
The Field of a Moving Charge 96
The Field of an Accelerating Charge 96
X-Ray Machines 98
The Universal Origin of Radiation 98
The Field of an Oscillating Charge 99
The Field of a Direct Current 99
The Field of an Alternating Current 102
Near and Far Field 105
The Fraunhoffer and Fresnel Zones 107
Parting Words 108

6 RELATIVITY AND QUANTUM PHYSICS 111
Relativity and Maxwell's Equations 111
Space and Time Are Relative 115

CONTENTS   vii

Space and Time Become Space-Time   120
The Cosmic Speed Limit and Proper Velocity   120
Electric Field and Magnetic Field Become the
  Electromagnetic Field   124
The Limits of Maxwell's Equations   125
Quantum Physics and the Birth of the Photon   126
The Quantum Vacuum and Virtual Photons   130
Explanation of the Magnetic Vector Potential   133
The Future of Electromagnetics   133
Relativity, Quantum Physics, and Beyond   134

7   THE HIDDEN SCHEMATIC   139
    The Non-Ideal Resistor   139
    The Non-Ideal Capacitor   142
    The Non-Ideal Inductor   143
    Non-Ideal Wires and Transmission Lines   146
    Other Components   149
    Making High-Frequency Measurements of Components   150
    RF Coupling and RF Chokes   150
    Component Selection Guide   151

8   TRANSMISSION LINES   153
    The Circuit Model   153
    Characteristic Impedance   155
    The Waveguide Model   157
    Relationship between the Models   159
    Reflections   159
    Putting It All Together   161
    Digital Signals and the Effects of Rise Time   163
    Analog Signals and the Effects of Frequency   165
    Impedance Transforming Properties   167
    Impedance Matching for Digital Systems   171
    Impedance Matching for RF Systems   172
    Maximum Load Power   173
    Measuring Characteristic Impedance: TDRs   175
    Standing Waves   177

9   WAVEGUIDES AND SHIELDS   181
    Reflection of Radiation at Material Boundaries   182
    The Skin Effect   183
    Shielding in the Far Field   184
    Near Field Shielding of Electric Fields   190
    Why You Should Always Ground a Shield   190

Near Field Shielding of Magnetic Fields 191
Waveguides 194
Resonant Cavities and Schumann Resonance 204
Fiber Optics 204
Lasers and Lamps 205

10 CIRCUITS AS GUIDES FOR WAVES AND S-PARAMETERS 209
Surface Waves 210
Surface Waves on Wires 213
Coupled Surface Waves and Transmission Lines 214
Lumped Element Circuits versus Distributed Circuits 217
$\lambda/8$ Transmission Lines 218
S-Parameters: A Technique for All Frequencies 219
The Vector Network Analyzer 223

11 ANTENNAS: HOW TO MAKE CIRCUITS THAT RADIATE 229
The Electric Dipole 229
The Electric Monopole 230
The Magnetic Dipole 230
Receiving Antennas and Reciprocity 231
Radiation Resistance of Dipole Antennas 231
Feeding Impedance and Antenna Matching 232
Antenna Pattern versus Electrical Length 236
Polarization 239
Effects of Ground on Dipoles 241
Wire Losses 244
Scattering by Antennas, Antenna Aperture, and Radar Cross-Section 245
Directed Antennas and the Yagi-Uda Array 246
Traveling Wave Antennas 246
Antennas in Parallel and the Folded Dipole 248
Multiturn Loop Antennas 249

12 EMC 251

Part I: Basics
Self-Compatibility and Signal Integrity 251
Frequency Spectrum of Digital Signals 252
Conducted versus Induced versus Radiated Interference 255
Crosstalk 257

Part II: PCB Techniques
Circuit Layout 259
PCB Transmission Lines 260

The Path of Least Impedance   262
The Fundamental Rule of Layout   264
Shielding on PCBs   265
Common Impedance: Ground Rise and Ground Bounce   267
Star Grounds for Low Frequency   269
Distributed Grounds for High Frequency: The 5/5 Rule   269
Tree or Hybrid Grounds   270
Power Supply Decoupling: Problems and Techniques   271
Power Supply Decoupling: The Design Process   278
RF Decoupling   282
Power Plane Ripples   282
90 Degree Turns and Chamfered Corners   282
Layout of Transmission Line Terminations   283
Routing of Signals: Ground Planes, Image Planes, and PCB Stackup   285
3W Rule for Preventing Crosstalk   286
Layout Miscellany   286
Layout Examples   287

Part III: Cabling
Ground Loops (Multiple Return Paths)   287
Differential Mode and Common Mode Radiation   290
Cable Shielding   296

13 LENSES, DISHES, AND ANTENNA ARRAYS   307
Reflecting Dishes   307
Lenses   311
Imaging   313
Electronic Imaging and Antenna Arrays   316
Optics and Nature   319

14 DIFFRACTION   321
Diffraction and Electrical Size   321
Huygens' Principle   323
Babinet's Principle   324
Fraunhofer and Fresnel Diffraction   325
Radio Propagation   326
Continuous Media   327

15 FREQUENCY DEPENDENCE OF MATERIALS, THERMAL RADIATION, AND NOISE   331
Frequency Dependence of Materials   331
Heat Radiation   338

Circuit Noise 343
Conventional and Microwave Ovens 343

APPENDIX A ELECTRICAL ENGINEERING BOOK
RECOMMENDATIONS 349

INDEX 353

PREFACE

This book is the result of many years of wondering about and researching the conceptual foundations of electromagnetics. My goal was to write a book that provided the reader with a conceptual understanding of electromagnetics and the insight to efficiently apply this understanding to real problems that confront scientists, engineers, and technicians. The fundamental equations that govern electromagnetic phenomena are those given to us by James Clerk Maxwell, and are commonly known as Maxwell's equations. Excepting quantum phenomena, all electromagnetic problems can be solved from Maxwell's equations. (The complete theory of electromagnetics, which includes quantum effects, is quantum electrodynamics, often abbreviated as QED.) However, many people lack the time and/or mathematical background to pursue the laborious calculations involved with the equations of electromagnetism. Furthermore, mathematics is just a tool, albeit a very powerful tool. For many problems, exacting calculations are not required. To truly understand, develop, and apply any branch of science requires a solid conceptual understanding of the material. As Albert Einstein stated, "Physics is essentially an intuitive and concrete science. Mathematics is only a means for expressing the laws that govern phenomena."* To this end, this book does not present Maxwell's equations and does not require any knowledge of these equations; nor is it required for the reader to know calculus or advanced mathematics.

The lack of advanced math in this book, I'm sure, will be a tremendous relief to most readers. However, to some readers, lack of mathematical rigor will be a negative attribute and perhaps a point for criticism. I contend that as long as the facts are correct and presented clearly, mathematics is not necessary for fundamental understanding, but rather for detailed treatment of problems. Moreover, everyday scientific practice shows that knowing the mathematical theory does not

---

*Quoted in A. P. French, ed., *Einstein: A Centenary Volume*, Cambridge, Mass.: Harvard University Press, 1979, p. 9.

ensure understanding of the real physical "picture." Certainly, mathematics is required for any new theories or conclusions. The material that I cover has been addressed formally in the literature, and readers are encouraged to pursue the numerous references given throughout. Conceptual methods for teaching the physical sciences have long been in use, but I think that the field of electromagnetics has been neglected and needs a book such as this. If relativity, quantum theory, and particle physics can be taught without mathematics, why not electromagnetics?

As inspiration and guide for my writing I looked to the style of writing in works such as *The Art of Electronics* by Paul Horowitz and Winfield Hill, several books by Richard Feynman, and the articles of the magazine *Scientific American*.

SUGGESTED AUDIENCE AND GUIDE FOR USE

This text is mainly intended as an introductory guide and reference for engineers and students who need to apply the concepts of electromagnetics to real-world problems in electrical engineering. Germane disciplines include radio frequency (RF) design, high-speed digital design, and electromagnetic compatibility (EMC). Electromagnetism is the theory that underlies all of electronics and circuit theory. With circuit theory being only an approximation, many problems, such as those of radiation and transmission line effects, require a working knowledge of electromagnetic concepts. I have included practical tips and examples of real applications of electromagnetic concepts to help the reader bridge the gap between theory and practice.

Taking a more general view, this book can be utilized by anyone learning electromagnetics or RF theory, be they scientist, engineer, or technician. In addition to self-study, it could serve well as a companion text for a traditional class on electromagnetics or as a companion text for classes on RF or high-speed electronics.

Those readers interested in RF or electromagnetics in general will find the entire book useful. While Chapter 1 serves as a good introduction for everyone, Chapters 2, 3, and 4 cover the basics and may be unnecessary for those who have some background in electromagnetics. I direct those readers whose discipline is digital design to focus on Chapters 1, 7, 8, and 12. These four chapters cover the important topics that relate to digital circuits and electromagnetic compatibility. EMC engineers should also focus on these four chapters, and in addition will probably be interested in the chapters that cover radiation (Chapter 5), shielding (Chapter 9), and antennas (Chapter 11). Chapter 6, which covers relativity and quantum theory, is probably not necessary for a book like this, but I have included it because these topics are fascinating to learn about and provide a different perspective of the electromagnetic field.

PARTING NOTES

I gladly welcome comments, corrections, and questions, as well as suggestions for topics of interest for possible future editions of this book. As with any writing endeavor, the publishing deadline forces the author to only briefly address some topics and omit some topics all together. I am also considering teaching one- or two-day professional courses covering selected material. Please contact me if such a course may be of interest to your organization. Lastly, I hope this book is as much a pleasure to read as it was to write.

Ron Schmitt, rfschmitt@ieee.org
Orono, Maine
July 2001

ACKNOWLEDGMENTS

First and foremost, I want to thank my wife, Kim Tripp. Not only did she give me love and patient support, she also typed in the references and drew many of the figures. For this, I am greatly indebted. I also want to thank my family, and particularly wish to thank my mother, Marion Schmitt, who provided the cover art and the drawings of hands and human figures in Chapter 3.

I am very thankful for the help of Dr. Laszlo Kish, for being a colleague and a friend, and most of all, for being my mentor. He had the patience to answer so many of my endless questions on electromagnetics, quantum physics, and physics in general. My bosses at SRD also deserve special mention: Mr. Carl Freeman, President; Dr. Greg Grillo, Vice President; and Dr. Jeremy Hammond, Director of Engineering Systems. Thanks to my friends at SRD for the most enjoyable years of my career.

This book wouldn't have been possible without the help of the great people at Newnes, particularly Candy Hall, Carrie Wagner, Chris Conty, Jennifer Packard, and Kevin Sullivan. Joan Lynch was instrumental to the success of this book by connecting me with Newnes. The readers of EDN, whose interest motivated me to write this book, deserve acknowledgment, as do my friends at Nortel Networks, where I wrote the first article that started this whole process.

Many people provided me with technical assistance in the writing. Roy McCammon pointed out that I didn't understand electromagnetics as well as I thought I did, especially in regard to surface waves in transmission lines. Dr. Keith Hardin provided me with his wonderful thesis on asymmetric currents and their relation to common-mode radiation. Dr. Clayton Paul examined my shielding plots and confirmed their correctness. Dr. Mark Rodwell provided me with insights on the state-of-the-art in ultra-high-speed electronics. Dr. Paul Horowitz told me about the strange problems involving cable braids at high frequencies. Dr. Thomas Jones and Dr. Jeremy Smallwood gave answers to questions regarding static electricity. Dr. Istvan Novak provided information on decoupling in high-speed digital systems. Dr. Allan Boardman answered several of my questions regarding electromagnetic surface waves. Dr. Tony Heinz helped me answer some questions regarding transmission lines in the infrared and beyond. I also wish to thank Nancy Lloyd, Daniel Starbird, and Julie Frost-Pettengill.

I want to thank all the people who reviewed my work: Don McCann, John Allen, Jesse Parks, Dr. Neil Comins, Les French, Dr. Fred Irons, Dr. Dwight Jaggard, and my anonymous reviewer at EDN. Finally, I extend thanks to everyone who made other small contributions and to anyone I may have forgotten in this list.

1 INTRODUCTION AND SURVEY OF THE ELECTROMAGNETIC SPECTRUM

How does electromagnetic theory tie together such broad phenomena as electronics, radio waves, and light? Explaining this question in the context of electronics design is the main goal of this book. The basic philosophy of this book is that by developing an understanding of the fundamental physics, you can develop an intuitive feel for how electromagnetic phenomena occur. Learning the physical foundations serves to build the confidence and skills to tackle real-world problems, whether you are an engineer, technician, or physicist.

The many facets of electromagnetics are due to how waves behave at different frequencies and how materials react in different ways to waves of different frequency. Quantum physics states that electromagnetic waves are composed of packets of energy called photons. At higher frequencies each photon has more energy. Photons of infrared, visible light, and higher frequencies have enough energy to affect the vibrational and rotational states of molecules and the electrons in orbit of atoms in the material. Photons of radio waves do not have enough energy to affect the bound electrons in a material. Furthermore, at low frequencies, when the wavelengths of the EM waves are very long compared to the dimensions of the circuits we are using, we can make many approximations leaving out many details. These low-frequency approximations give us the familiar world of basic circuit theory.

THE NEED FOR ELECTROMAGNETICS

So why would an electrical engineer need to know all this theory? There are many reasons why any and all electrical engineers need to understand electromagnetics. Electromagnetics is necessary for achieving electromagnetic compatibility of products, for understanding high-speed digital electronics, RF, and wireless, and for optical computer networking.

Certainly any product has some electromagnetic compatibility (EMC) requirements, whether due to government mandated standards or simply for the product to function properly in the intended environment. In most EMC problems, the product can be categorized as either an aggressor or a victim. When a product is acting as an aggressor, it is either radiating energy or creating stray reactive fields at power levels high enough to interfere with other equipment. When a product is acting as a victim, it is malfunctioning due to interference from other equipment or due to ambient fields in its environment. In EMC, victims are not always blameless. Poor circuit design or layout can create products that are very sensitive to ambient fields and susceptible to picking up noise. In addition to aggressor/victim problems, there are other problems in which noise disrupts proper product operation. A common problem is that of cabling, that is, how to bring signals in and out of a product without also bringing in noise and interference. Cabling problems are especially troublesome to designers of analog instrumentation equipment, where accurately measuring an external signal is the goal of the product.

Moreover, with computers and networking equipment of the 21st century running at such high frequencies, digital designs are now in the RF and microwave portion of the spectrum. It is now crucial for digital designers to understand electromagnetic fields, radiation, and transmission lines. This knowledge is necessary for maintaining signal integrity and for achieving EMC compliance. High-speed digital signals radiate more easily, which can cause interference with nearby equipment. High-speed signals also more often cause circuits within the same design to interfere with one another (i.e., crosstalk). Circuit traces can no longer be considered as ideal short circuits. Instead, every trace should be considered as a transmission line because reflections on long traces can distort the digital waveforms. The Internet and the never-ending quest for higher bandwidth are pushing the speed of digital designs higher and higher. Web commerce and applications such as streaming audio and video will continue to increase consumer demand for higher bandwidth. Likewise, data traffic and audio and video conferencing will do the same for businesses. As we enter the realm of higher frequencies, digital designs are no longer a matter of just ones and zeros.

Understanding electromagnetics is vitally important for RF (radio frequency) design, where the approximations of electrical circuit theory start to break down. Traditional viewpoints of electronics (electrons flowing in circuits like water in a pipe) are no longer sufficient for RF designs. RF design has long been considered a "black art," but it is time to put that myth to rest. Although RF design is quite different from low-frequency design, it is not very hard to understand for any electrical engineer. Once you understand the basic concepts and gain an intuition for how electromagnetic waves and fields behave, the mystery disappears.

Optics has become essential to communication networks. Fiber optics are already the backbone of telecommunications and data networks. As we exhaust the speed limits of electronics, optical interconnects and possibly optical computing will start to replace electronic designs. Optical techniques can work at high speeds and are well suited to parallel operations, providing possibilities for computation rates that are orders of magnitude faster than electronic computers. As the digital age progresses, many of us will become "light engineers," working in the world of photonics. Certainly optics is a field that will continue to grow.

THE ELECTROMAGNETIC SPECTRUM

For electrical engineers the word *electromagnetics* typically conjures up thoughts of antennas, transmission lines, and radio waves, or maybe boring lectures and "all-nighters" studying for exams. However, this electrical word also describes a broad range of phenomena in addition to electronics, ranging from X-rays to optics to thermal radiation. In physics courses, we are taught that all these phenomena concern electromagnetic waves. Even many nontechnical people are familiar with this concept and with the electromagnetic spectrum, which spans from electronics and radio frequencies through infrared, visible light, and then on to ultraviolet and X-rays. We are told that these waves are all the same except for frequency. However, most engineers find that even after taking many physics and engineering courses, it is still difficult to see much commonality across the electromagnetic spectrum other than the fact that all are waves and are governed by the same mathematics (Maxwell's equations). Why is visible light so different from radio waves? I certainly have never encountered electrical circuits or antennas for visible light. The idea seems absurd. Conversely, I have never seen FM radio or TV band lenses for sale. So why do light waves and radio waves behave so differently?

Of course the short answer is that it all depends on frequency, but on its own this statement is of little utility. Here is an analogy. From basic chemistry, we all know that all matter is made of atoms, and that atoms contain a nucleus of protons and neutrons with orbiting electrons. The characteristics of each element just depend on how many protons the atom has. Although this statement is illuminating, just knowing the number of protons in an atom doesn't provide much more than a framework for learning about chemistry. Continuing this analogy, the electromagnetic spectrum as shown in Figure 1.1 provides a basic framework for understanding electromagnetic waves, but there is a lot more to learn.

To truly understand electromagnetics, it is important to view different problems in different ways. For any given frequency of a wave, there is also a corresponding wavelength, time period, and quantum of energy. Their definitions are given below, with their corresponding relationships in free space.

---

frequency, f, *the number of oscillations per second*
wavelength, λ, *the distance between peaks of a wave*:

$$\lambda = \frac{c}{f}$$

time period, T, *the time between peaks of a wave*:

$$T = \frac{1}{f}$$

photon energy, E, *the minimum value of energy that can be transferred at this frequency*:

$$E = h \times f$$

where c equals the speed of light and h is Planck's constant.

Depending on the application, one of these four interrelated values is probably more useful than the others. When analyzing digital transmission lines, it helps to compare the signal rise time to the signal transit time down the transmission line. For antennas, it is usually most intuitive to compare the wavelength of the signal to the antenna length. When examining the resonances and relaxation of dielectric materials it helps to compare the frequency of the waves to the resonant frequency of the material's microscopic dipoles. When dealing with infrared, optical, ultraviolet, and X-ray interactions with matter, it is often most useful to talk about the energy of each photon to relate it to the orbital energy of electrons in atoms. Table 1.1 lists these four values at various

THE ELECTROMAGNETIC SPECTRUM

Figure 1.1   The electromagnetic spectrum.

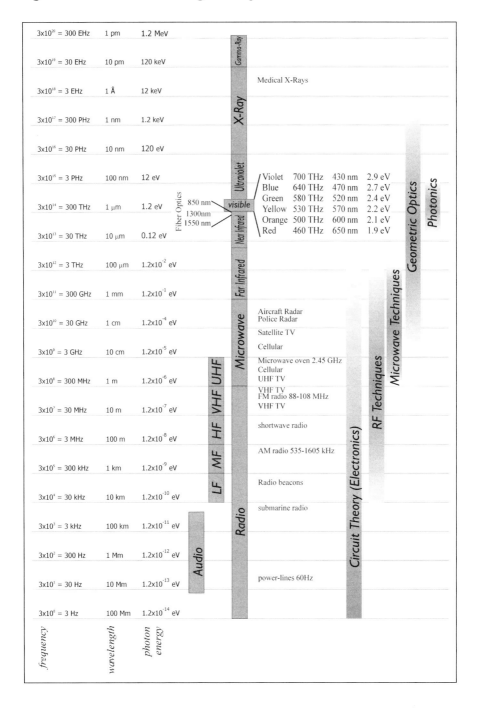

Table 1.1 Characteristics of Electromagnetic Waves at Various Frequencies

Frequency	Wavelength	Photon Energy	Period	Copper Skin Depth	Copper Propagation Phase Angle
60 Hz Power line frequency	5000 km	2.48×10^{-13} eV	16.7 msec	8.4 mm	45° (conductor)
440 Hz audio	681 km	1.82×10^{-12} eV	2.27 msec	3.1 mm	45° (conductor)
1 MHz AM radio	300 km	4.14×10^{-9} eV	1.00 μsec	65 μm	45° (conductor)
100 MHz FM radio	3.00 m	4.14×10^{-7} eV	10.0 nsec	6.5 μm	45° (conductor)
2.45 GHz Microwave oven	12.2 cm	1.01×10^{-7} eV	40.8 psec	1.3 μm	45° (conductor)
160 GHz Cosmic background radiation ("Big Bang") peak	1.87 mm	6.62×10^{-4} eV	6.25 psec	0.16 μm	46° (conductor)
4.7 THz Relaxation resonance of copper	63.8 μm	1.94×10^{-2} eV	213 fsec	27.3 nm	68°
17.2 THz Room temperature Blackbody infrared peak	17.4 μm	7.11×10^{-2} eV	5.81 fsec	21.8 nm	82°
540 THz Center of visible band	555 nm	2.23 eV	1.85 fsec	21.8 nm	90° (reflecting plasma)
5000 THz Ultraviolet	60.0 nm	20.7 eV	0.60 fsec	89 μm	0° (transparent plasma)
1×10^7 THz Diagnostic x-ray	30 pm	4.14×10^4 eV	1.00×10^{-19} sec	400 m	0° (transparent plasma)
1×10^8 THz Gamma ray from ^{198}Hg nucleus	3.0 pm	4.15×10^5 eV	1.00×10^{-20} sec	40 km	0° (transparent plasma)

THE ELECTROMAGNETIC SPECTRUM

Dipole Radiation Field Border	Blackbody Characteristic Radiation Temperature	Photon Rate for 1 mW Source	Aperture for Human Quality Imaging	Aperture for Minimal Quality Imaging
795 km	<1°K	2.5×10^{28} photons/sec	2.7×10^{10} m	7.0×10^{7} m
108 km	<1°K	3.4×10^{27} photons/sec	3.7×10^{9} m	9.5×10^{6} m
47.7 m	<1°K	1.5×10^{24} photons/sec	1.6×10^{6} m	4200 m
47.7 cm	<1°K	1.5×10^{22} photons/sec	1600 m	42 m
1.95 cm	<1°K	6.2×10^{20} photons/sec	660 m	1.7 m
298 μm	2.72°K (temperature of outer space)	9.4×10^{18} photons/sec	10 m	2.6 cm
40.2 μm	80°K	3.2×10^{17} photons/sec	35 cm	0.89 mm
2.77 μm	20°C	8.8×10^{16} photons/sec	9.4 cm	0.24 mm
88.4 nm	9440°K	2.8×10^{15} photons/sec	3.0 mm	7.8 μm
9.54 nm	85,000°K	3.0×10^{14} photons/sec	0.32 mm	840 nm
4.77 pm	1.7×10^{8}°K	1.5×10^{11} photons/sec	160 nm	420 pm
0.477 pm	1.7×10^{9}°K	1.5×10^{10} photons/sec	16 nm	42 pm

parts of the electromagnetic spectrum, and also includes some other relevant information. If some of these terms are unfamiliar to you, don't fret—they'll be explained as you progress through the book.

ELECTRICAL LENGTH

An important concept to aid understanding of electromagnetics is electrical length. Electrical length is a unitless measure that refers to the length of a wire or device at a certain frequency. It is defined as the ratio of the physical length of the device to the wavelength of the signal frequency:

$$\text{Electrical length} = \frac{L}{\lambda}$$

As an example, consider a 1-meter long antenna. At 1 kHz this antenna has an electrical length of about 3×10^{-6}. An equivalent way to say this is in units of wavelength; that is, a 1 meter antenna is $3 \times 10^{-6} \lambda$ long at 1 kHz. At 1 kHz this antenna is electrically short. However, at 100 MHz, the frequency of FM radio, this antenna has an electrical length of 0.3 and is considered electrically long. In general, any device whose electrical length is less than about 1/20 can be considered electrically short. (Beware: When working with wires that have considerable loss or large impedance mismatches, even electrical lengths of 1/50 may not be electrically short.) Circuits that are electrically short can in general be fully described by basic circuit theory without any need to understand electromagnetics. On the other hand, circuits that are electrically long require RF techniques and knowledge of electromagnetics.

At audio frequencies and below (<20 kHz), electromagnetic waves have very long wavelengths. The wavelength is typically much larger than the length of any of the wires in the circuit used. (An exception would be long telephone lines.) *When the wavelength is much longer than the wire lengths, the basic rules of electronic circuits apply and electromagnetic theory is not necessary.*

THE FINITE SPEED OF LIGHT

Another way of looking at low-frequency circuitry is that the period (the inverse of frequency) of the waves is much larger than the delay through the wires. "What delay in the wires?" you might ask. When we are involved in low-frequency circuit design it is easy to forget that the electrical signals are carried by waves and that they must travel at the speed of light, which is very fast (about 1 foot/nsec on open air wires), but not infinite. So, even when you turn on a light switch there is a delay before the light bulb receives the voltage. The same delay occurs between your home stereo and its speakers. This delay is typically too small for humans to perceive, and is ignored whenever you approximate a wire as an ideal short circuit. The speed of light delay also occurs in telephone lines, which can produce noticeable echo (>50 msec) if the connection spans a large portion of the earth or if a satellite feed is used. Long distance carriers use echo-cancellation electronics for international calls to suppress the effects. The speed of light delay becomes very important when RF or high-speed circuits are being designed. For example, when you are designing a digital system with 2 nsec rise-times, a couple feet of cable amounts to a large delay.

ELECTRONICS

Electronics is the science and engineering of systems and equipment that utilize the flow of electrons. Electrons are small, negatively charged particles that are free to move about inside conductors such as copper and gold. Because the free electrons are so plentiful inside a conductor, we can often approximate electron flow as fluid flow. In fact, most of us are introduced to electronics using the analogy of (laminar) flow of water through a pipe. Water pressure is analogous to electrical voltage, and water flow rate is analogous to electrical current. Frictional losses in the pipe are analogous to electrical resistance. The pressure drop in a pipe is proportional to the flow rate multiplied by the frictional constant of the pipe. In electrical terms, this result is Ohm's law. That is, the voltage drop across a device is equal to the current passing through the device multiplied by the resistance of the device:

$$\text{Ohm's law: } V = I \cdot R$$

Now imagine a pump that takes water and forces it through a pipe and then eventually returns the water back to the tank. The water in the tank is considered to be at zero potential—analogous to an electrical ground or common. A pump is connected to the water tank. The pump produces a pressure increase, which causes water to flow. The pump is like a voltage source. The water flows through the pipes, where frictional losses cause the pressure to drop back to the original "pressure potential." The water then returns to the tank. From the perspective of energy

Figure 1.2 A simple circuit demonstrating Kirchhoff's voltage law ($V = V_1 + V_2 + V_3$).

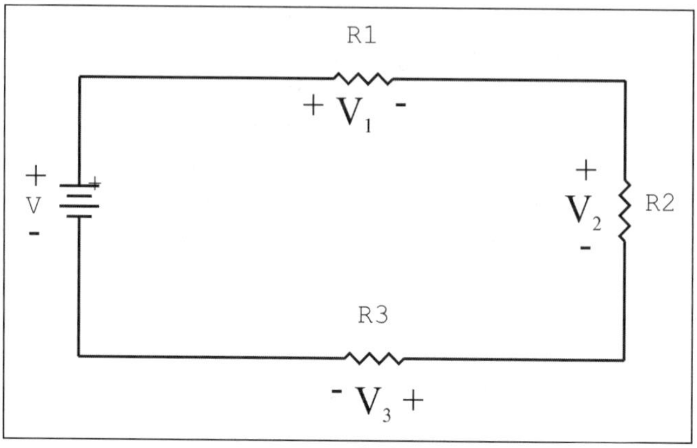

flow, the pump sources energy to the water, and then in the pipes all of the energy is lost due to friction, converted to heat in the process. Keep in mind that this analogy is only an approximation, even at DC.

Basic circuit theory can be thought of in the same manner. The current flows in a loop, or circuit, and is governed by Kirchhoff's laws (as shown in Figures 1.2 and 1.3). Kirchhoff's voltage law (KVL) says that the voltages in any loop sum to zero. In other words, for every voltage drop in a circuit there must be a corresponding voltage source. Current flows in a circle, and the total of all the voltage sources in the circle or circuit is always equal to the total of all the voltage sinks (resistors, capacitors, motors, etc.). KVL is basically a consequence of the conservation of energy.

Kirchhoff's current law (KCL) states that when two or more branches of a circuit meet, the total current is equal to zero. This is just conservation of current. For example, if 5 amps is coming into a node through a wire, then 5 amps must exit the node through another wire(s). In our water tank analogy, this law implies that no water can leave the system. Current can't just appear or disappear.

Additional rules of basic circuit theory are that circuit elements are connected through ideal wires. Wires are considered perfect conductors with no voltage drop or delay. The wires between components are therefore all considered to be at the same voltage potential and are referred to as a node. This concept often confuses the beginning student of electronics. For an example, refer to Figure 1.4. In most schematic diagrams, the wire connections are in fact considered to be ideal. This method of representing electronic circuits is termed "lumped element" design.

ELECTRONICS

Figure 1.3 A simple circuit demonstrating Kirchhoff's current law ($I_1 = I_2 + I_3$).

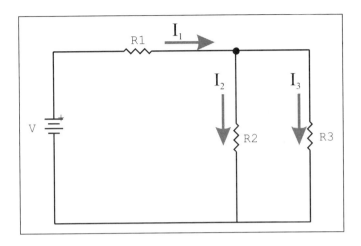

Figure 1.4 A simple circuit demonstrating the voltage node principle. The voltage is the same everywhere inside each of the dotted outlines.

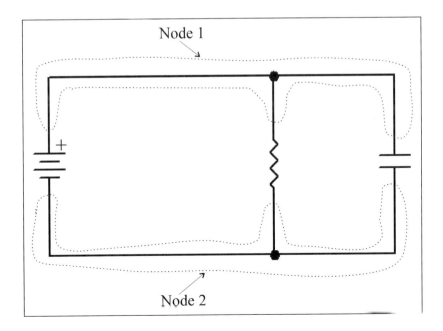

The ironic thing about this is that the beginning student is taught to ignore the shape and length of wires, but at RF frequencies the length and shape of the wires become just as important as the components. Engineering and science are filled with similar situations where you must develop a simplified understanding of things before learning all the exceptions and details. Extending the resistance concept to the concept of AC (alternating current) impedance allows you to include capacitors and inductors. That is circuit theory in a nutshell. There are no antennas or transmission lines. We can think of the circuit as electrons flowing through wires like water flowing through a pipe. Electromagnetics is not needed.

ANALOG AND DIGITAL SIGNALS

Electronics is typically divided into the categories of analog and digital. Analog signals are continuously varying signals such as audio signals. Analog signals typically occupy a specific bandwidth and can be decomposed in terms of sinusoids using Fourier theory. For example, signals carrying human voice signals through the telephone network occupy the frequency band from about 100 Hz to about 4000 Hz.

Digital signals, on the other hand, are a series of ones and zeroes. A typical method to represent a digital signal is to use 5 V for a one and 0 V for a zero. A digital clock signal is shown as an example in Figure 1.5. Fourier theory allows us to create such a square wave by summing individual sine waves. The individual sine waves are at multiples or harmonics of the clock frequency.* To create a perfectly square signal (signal rise and fall times of zero) requires an infinite number of harmonics, spanning to infinite frequency. Of course, this is impossible in reality, so all real digital signals must have rise and fall times greater than zero. In other words, no real digital signal is perfectly square. *When performing transmission line and radiation analysis for digital designs, the rise and fall times are the crucial parameters.*

RF TECHNIQUES

At higher frequencies, basic circuit theory runs into problems. For example, if wires are electrically long, transmission line effects can occur.

*Rock musicians may find it interesting to know that the signal of an electric guitar with distortion looks very similar to Figure 1.5. The distortion effect for guitars is created by "squaring off" the sine waves from the guitar, using a saturated amplifier.

RF TECHNIQUES

Figure 1.5 A 5 Hz clock signal and its frequency content.

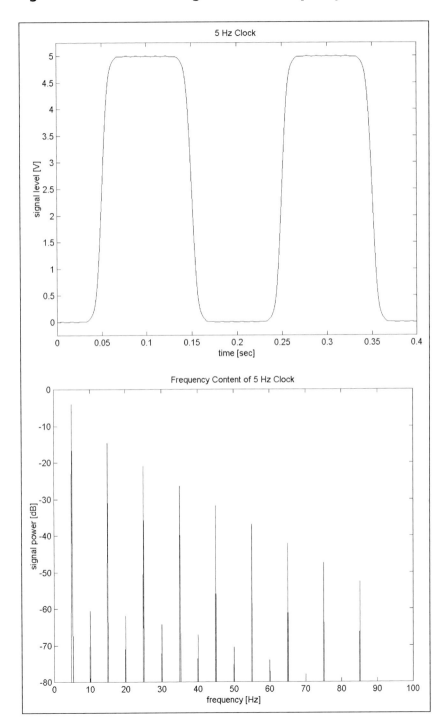

The basic theory no longer applies because electromagnetic wave reflections bouncing back and forth along the wires cause problems. These electromagnetic wave reflections can cause constructive or destructive interference resulting in the breakdown of basic circuit theory. In fact, when a transmission line has a length equal to one quarter wavelength of the signal, a short placed at the end will appear as an open circuit at the other end! Certainly, effects like this cannot be ignored. Furthermore, at higher frequencies, circuits can radiate energy much more readily; that is circuits can turn into antennas. Parasitic capacitances and inductances can cause problems too. No component can ever be truly ideal. The small inductance of component leads and wires can cause significant voltage drops at high frequencies, and stray capacitances between the leads of the component packages can affect the operation of a high-frequency circuit. These parasitic elements are sometimes called "the hidden schematic" because they typically are not included on the schematic symbol. (The high-frequency effects just mentioned are illustrated in Figure 1.6.)

How do you define the high-frequency regime? There is no exact border, but when the wavelengths of the signals are similar in size or smaller than the wire lengths, high-frequency effects become important; in other words, when a wire or circuit element becomes electrically long, you are dealing with the high-frequency regime. An equivalent way to state this is that when the signal period is comparable in magnitude or smaller than the delay through the interconnecting wires, high-frequency effects become apparent. *It is important to note that for digital signals, the designer must compare the rise and fall times of the digital signal to the wire delay.* For example, a 10 MHz digital clock signal may only have a signal period of 100 nsec, but its rise time may be as low as 5 nsec. Hence, the RF regime doesn't signify a specific frequency range, but signifies frequencies where the rules of basic circuit theory breakdown. *A good rule of thumb is that when the electrical length of a circuit element reaches 1/20, RF (or high-speed digital) techniques may need to be used.*

When working with RF and high-frequency electronics it is important to have an understanding of electromagnetics. At these higher frequencies, you must understand that the analogy of electrons acting like water through a pipe is really more of a myth than a reality. In truth, circuits are characterized by metal conductors (wires) that serve to guide electromagnetic energy. The circuit energy (and therefore the signal) is carried between the wires, and not inside the wires. For an example, consider the power transmission lines that deliver the electricity to our homes at 60 Hz. The electrons in the wires do not directly transport the energy from the power plant to our homes. On the contrary, the energy

Figure 1.6 Some effects that occur in high-frequency circuits.

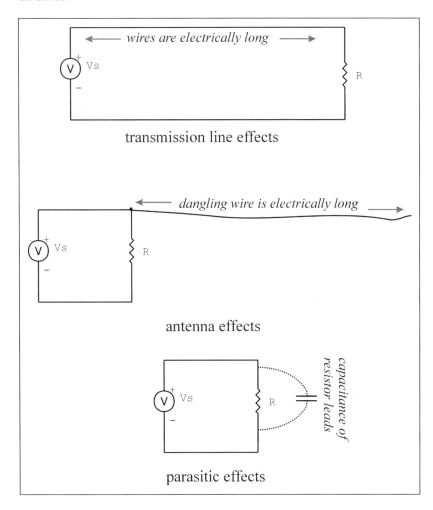

is carried in the electromagnetic field between the wires. This fact is often confusing and hard to accept for circuit designers. The wire electrons are not experiencing any net movement. They just slosh back and forth, and through this movement they propagate the field energy down the wires. A good analogy is a "bucket brigade" that people sometimes use to fight fires. A line of firefighters (analogous to the electrons) is set up between the water source (signal source) and the fire (the load). Buckets of water (the electromagnetic signal) are passed along the line from firefighter to firefighter. The water is what puts out the fire. The people are just there to pass the water along. In a similar manner, the electrons just serve to pass the electromagnetic signal from source to load. This statement is true at all electronic frequencies, DC, low frequency, and RF.

MICROWAVE TECHNIQUES

At microwave frequencies in the GHz range, circuit theory is no longer very useful at all. Instead of thinking about circuits as electrons flowing through a pipe, it is more useful to think about circuits as structures to guide and couple waves. At these high frequencies, lumped elements such as resistors, capacitors, and inductors are often not viable. As an example, the free space wavelength of a 30 GHz signal is 1 cm. Therefore, even the components themselves are electrically long and do not behave as intended. Voltage, current, and impedance are typically not used. In this realm, electronics starts to become similar to optics in that we often talk of power transmitted and reflected instead of voltage and current. Instead of impedance, reflection/transmission coefficients and S-parameters are used to describe electronic components. Some microwave techniques are shown in Figure 1.7.

INFRARED AND THE ELECTRONIC SPEED LIMIT

The infrared region is where the spectrum transitions from electronics to optics. The lower-frequency portion of the infrared is termed the "far infrared," and is the extension of the microwave region. Originally, the edge of the microwave band (300 MHz) was considered the highest viable frequency for electronics. As technology progresses, the limit of electronics extends further into the infrared. Wavelengths in the infrared are under 1 mm, implying that even a 1 mm wire is electrically long, readily radiating energy from electrical currents. Small devices are therefore mandatory.

At the time of publishing of this book, experimental integrated circuit devices of several terahertz (10^{12} Hz) had been achieved, and 40 GHz digital devices had become commercially available for communications applications. (Terahertz devices were created decades ago using vacuum tube techniques, but these devices are obviously not viable for computing devices.) Certainly digital devices in the hundreds of gigahertz will become commercially viable; in fact, such devices have already been demonstrated by researchers. Making digital devices past terahertz speeds will be a very difficult challenge. To produce digital waveforms,

Figure 1.7 Examples of microwave techniques.

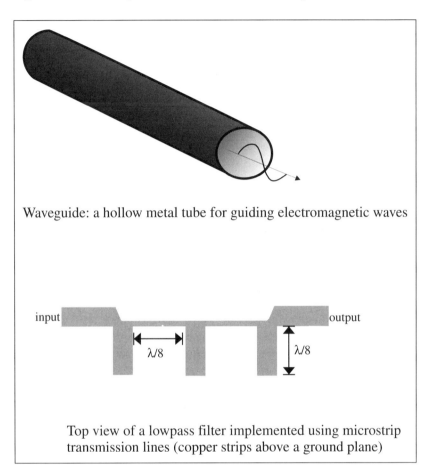

Waveguide: a hollow metal tube for guiding electromagnetic waves

Top view of a lowpass filter implemented using microstrip transmission lines (copper strips above a ground plane)

you need an amplifier with a bandwidth of at least 3 to 5 times the clock frequency. Already researchers are pursuing special semiconductors such as Indium Phosphide (InP) electron spin, single-electron, and quantum devices, as well as molecular electronics. Only time will tell what the ultimate "speed limit" for electronics will be.

What is almost certain is that somewhere in the infrared frequencies, electronics will always be impossible to design. There are many problems in the infrared facing electronics designers. The speed of transistors is limited by their size; consequently, to probe higher frequencies, the state of the art in integrated circuit geometries must be pushed to smaller and smaller sizes. Quantum effects, such as tunneling, also cause problems. Quantum tunneling allows electrons to pass through the gate of very small MOSFET transistors. This effect is a major problem facing researchers trying to further shrink CMOS technology. Furthermore, the properties of most materials begin to change in the infrared. The conductive properties of metals begin to change. In addition, most dielectric materials become very lossy. Even dielectrics that are transparent in the visible region, such as water and glass, become opaque in the portions of the infrared. Photons in the infrared are very energetic compared to photons at radio frequencies and below. Consequently, infrared photons can excite resonant frequencies in materials. Another characteristic of the infrared is that the maximum of heat radiation occurs in the infrared for materials between room temperature (20°C) and several thousand degrees Celsius. These characteristics cause materials to readily absorb and emit radiation in the infrared. For these reasons, we can readily feel infrared radiation. The heat we feel from incandescent lamps is mostly infrared radiation. It is absorbed very easily by our bodies.

VISIBLE LIGHT AND BEYOND

At the frequencies of visible light, many dielectrics become less lossy again. Materials such as water and glass that are virtually lossless with respect to visible light are therefore transparent. Considering that our eyes consist mostly of water, we are very fortunate that water is visibly transparent. Otherwise, our eyes, including the lens, would be opaque and quite useless. A striking fact of nature is that the absorption coefficient of water rises more than 7 decades (a factor of 10 million) in magnitude on either side of the visible band. So it is impossible to create a reasonably sized, water-based eye at any other part of the spectrum. All creatures with vision exploit this narrow region of the spectrum. Nature is quite amazing!

At visible frequencies, the approximations of geometric optics can be used. These approximations become valid when the objects used become much larger than a wavelength. This frequency extreme is the opposite of the circuit theory approximations. The approximation is usually called ray theory because light can be approximated by rays or streams of particles. Isaac Newton was instrumental in the development of geometric optics, and he strongly argued that light consisted of particles and not waves. The physicist Huygens developed the wave theory of light and eventually experimental evidence proved that Huygens was correct. However, for geometrical optics, Newton's theory of particle streams works quite well. An example of geometrical optics is the use of a lens to concentrate or focus light. Figure 1.8 pro-

Figure 1.8 A lens that focuses rays of light.

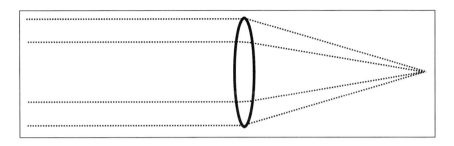

vides a lens example. Most visible phenomena, including our vision, can be studied with geometrical optics. The wave theory of light is usually needed only when studying diffraction (bending of light around corners) and coherent light (the basis for lasers). Wave theory is also needed to explain the resolution limits of optical imaging systems. A microscope using visible light can only resolve objects down to about the size of a wavelength.

At the range of ultraviolet frequencies and above (X-rays, etc.) each photon becomes so energetic that it can kick electrons out of their atomic orbit. The electron becomes free and the atom becomes ionized. Molecules that absorb these high-energy photons can lose the electrons that bond the molecules together. Ions and highly reactive molecules called *free radicals* are produced. These highly reactive ions and molecules cause cellular changes and lead to biological tissue damage and cancer. Photons of visible and infrared light, on the other hand, are less energetic and only cause molecular heating. We feel the heat of the infrared radiation from the sun. We see the light of visible radiation from the sun. Our skin is burned and damaged by the ultraviolet radiation from the sun.

X-ray photons, being higher in energy, are even more damaging. Most materials are to some degree transparent to X-rays, allowing the use of X-ray photography to "see through" objects. But when X-rays are absorbed, they cause cellular damage. For this reason, limited X-ray exposure is recommended by physicians. The wavelengths of high-energy X-rays are about the same size as the atomic spacing in matter. Therefore, to X-rays, matter cannot be approximated as continuous, but rather is "seen" as lumps of discrete atoms. The small wavelength makes X-rays useful for studying crystals such as silicon, using the effects of diffraction. Above X-rays in energy are gamma rays and cosmic rays. These extremely high-energy waves are produced only in high-energy phenomena such as radioactive decay, particle physics collisions, nuclear power plants, atomic bombs, and stars.

LASERS AND PHOTONICS

Electronic circuits can be created to transmit, amplify, and filter signals. These signals can be digital bits or analog signals such as music or voices. The desire to push electronics to higher frequencies is driven by two main applications: computers and communication links. For computers, higher frequencies translate to faster performance. For communication links, higher frequencies translate to higher bandwidth. Oscillator circuits serve as timing for both applications. Computers are in general synchronous and require a clock signal. Communications links need a carrier signal to modulate the information for transmission. Therefore, a basic need to progress electronics is the ability to create oscillators.

In the past few decades, photonics has emerged as an alternative to electronics, mostly in communication systems. Lasers and fiber optic cables are used to create and transmit pulses of a single wavelength (frequency) of light. In the parlance of optics, single-frequency sources are known as coherent sources. Lasers produce synchronized or coherent photons; hence, the name photonics. The light that we encounter every day from the sun and lamps is noncoherent light. If we could look at this light on an oscilloscope, it would look like noise. In fact, the visible light that we utilize for our vision is noise—the thermal noise of hot objects such as the sun or the filament in a light bulb. The electrical term "white noise" comes from the fact that optical noise contains all the visible colors (frequencies) and appears white. The white noise of a light bulb extends down to electronic frequencies and is the same white noise produced by resistors and inherent in all circuits. Most imaging devices, like our eyes and cameras, only use the average squared-field amplitude of the light received. (Examination at the quantum level reveals imaging devices to be photon detectors/counters.) Averaging allows us to use "noisy" signals for vision, but because of averaging all phase information is lost. To create sophisticated communication devices, such light is not suitable. Instead the coherent, single-frequency light of lasers is used. Lasers make high-bandwidth fiber optic communication possible.

Until recently, the major limitation of photonics was that the laser pulsed signals eventually had to be converted to electronic signals for any sort of processing. For instance, in data communications equipment, major functions include the switching, multiplexing, and routing of data between cables. In the past, only electronic signals could perform these functions. This requirement limited the bandwidth of a fiber optic cable to the maximum available electronic bandwidth. However, with recent advances in optical multiplexing and switching, many tasks can now be performed completely using photonics. The upshot has been an exponential increase in the data rates that can be achieved with fiber optic technology. The ultimate goal for fiber optics communication is to create equipment that can route Internet protocol (IP) datapackets using only photonics. Such technology would also lead the way for optical computing, which could provide tremendous processing speeds as compared with electronic computers of today.

SUMMARY

Different techniques and approximations are used in the various portions of the electromagnetic spectrum. Basic circuit theory is an approximation made for low-frequency electronics. The circuit theory approximations work when circuits are electrically small. In other words, circuit theory is the limit of electromagnetics as the wavelength becomes infinitely larger than the circuit. RF theory takes circuit theory and adds in some concepts and relations from electromagnetics. RF circuit theory accounts for transmission line effects in wires and for antenna radiation. At microwave frequencies it becomes impossible to design circuits with lumped elements like resistors, capacitors, and inductors because the wavelengths are so small. Distributed techniques must be used to guide and process the waves. In the infrared region, we can no longer design circuits. The wavelengths are excessively small, active elements like transistors are not possible, and most materials become lossy, readily absorbing and radiating any electromagnetic energy. At the frequencies of visible light, the wavelengths are typically much smaller than everyday objects, and smaller than the human eye can notice. In this range, the approximations of geometrical optics are used. Geometrical optics is the limit of electromagnetic theory where wavelength becomes infinitely smaller than the devices used. At frequencies above light, the individual photons are highly energetic, able to break molecular bonds and cause tissue damage.

With the arrival of the information age, we rely on networked communications more and more every day, from our cell phones and pagers to our high-speed local-area networks (LANs) and Internet connections. The hunger for more bandwidth consistently pushes the frequency and complexity of designs. The common factor in all these applications is that they require a good understanding of electromagnetics.

BIBLIOGRAPHY: GENERAL TOPICS FOR CHAPTER 1

Button, K. J., Editor, *Infrared and Millimeter Waves, Volume I: Sources of Radiation*, New York: Academic Press, 1979.

Cogdell, J. R., *Foundations of Electrical Engineering*, 2nd Edition, Englewood Cliffs, NJ: Prentice-Hall, 1995.

Encyclopedia Britannica Inc., "Electromagnetic Radiation," "Laser," *Encyclopedia Britannica*, Chicago: Encyclopedia Britannica Inc., 1999.

Feynman, R. P., R. B. Leighton, M. Sands, *The Feynman Lectures on Physics Vol I: Mainly Mechanics, Radiation, and Heat*, Reading, Mass.: Addison-Wesley Publishing, 1963.

Feynman, R. P., R. B. Leighton, M. Sands, *The Feynman Lectures on Physics Vol II: Mainly Electromagnetism and Matter*, Reading, Mass.: Addison-Wesley Publishing, 1964.

Granatstein, V. L., and I. Alexeff, Editors, *High-Power Microwave Sources*, Boston: Artech House, 1987.

Halliday, D., R. Resnick, J. Walker, *Fundamentals of Physics*, 6th Edition, New York: John Wiley & Sons, 2000.

Halsall, F., *Data Communications, Computer Networks and Open Systems*, 4th Edition, Reading, Mass.: Addison-Wesley, 1996.

Halsall, F., *Multimedia Communications: Applications, Networks, Protocols, and Standards*, Reading, Mass.: Addison-Wesley, 2000.

Hecht, E., and K. Guardino, *Optics*, 3rd Edition, Reading, Mass.: Addison-Wesley, 1997.

Hutchinson, C., J. Kleinman, D. R. Straw, Editors, *The ARRL Handbook for Radio Amateurs*, 78th edition, Newington, Conn.: American Radio Relay League, 2001.

Johnson, H., and M. Graham, *High-Speed Digital Design: A Handbook of Black Magic*, Englewood Cliffs, NJ: Prentice-Hall, 1993.

Kraus, J. D., and D. A. Fleisch, *Electromagnetics with Applications*, 5th Edition, Boston: McGraw-Hill, 1999.

Montrose, M. I., *Printed Circuit Board Design Techniques EMC Compliance—A Handbook for Designers*, 2nd Edition, New York: IEEE Press, 2000.

Paul, C. R., *Introduction to Electromagnetic Compatibility*, New York: John Wiley & Sons, 1992.

Pedrotti, F. L., and L. S. Pedrotti, *Introduction to Optics*, 2nd Edition, Upper Saddle River, NJ: Prentice Hall, 1993.

Pozar, D. M., *Microwave Engineering*, 2nd Edition, New York: John Wiley, 1998.

Schmitt, R., "Analyze Transmission Lines with (almost) No Math", *EDN*, March 18, 1999.

Schmitt, R., "Understanding Electromagnetic Fields and Antenna Radiation Takes (almost) No Math", *EDN*, March 2, 2000.

Straw, R. D., Editor, *The ARRL Antenna Book*, 19th Edition, Newington, Conn.: American Radio Relay League, 2000.

Tanenbaum, S., *Computer Networks*, 3rd Edition, Upper Saddle River, NJ: Prentice Hall, 1996.

BIBLIOGRAPHY: STATE-OF-THE-ART ELECTRONICS

Brock, D. K., E. K. Track, J. M. Rowell, "Superconductor ICs: The 100-GHz Second Generation," *IEEE Spectrum*, December 2000.

Collins, P. G., and P. Avouris, "Nanotubes for Electronics," *Scientific American*, December 2000.

Cravotta, N., "DWDM: Feeding Our Insatiable Appetite for Bandwidth," *EDN*, September 1, 2000.

Geppert, L., "Quantum Transistors: Toward Nanoelectronics," *IEEE Spectrum*, September 2000.

Hopkins, J.-M., and W. Sibbett, "Big Payoffs in a Flash," *Scientific American*, September 2000.

Israelsohn, J., "Switching the Light Fantastic," *EDN*, October 26, 2000.

Israelsohn, J., "Pumping Data at Gigabit Rates," *EDN*, April 12, 2001.

Matsumoto, C., and L. Wirbel, "Vitesse goes with InP process for 40-Gbit devices," *EETimes.com*, CMP Media Inc. 2000.

Mullins, J., "The Topsy Turvy World of Quantum Computing," *IEEE Spectrum*, February 2001.

Nortel Networks, "Pushing the Limits of Real-World Optical Networks," *Nortel's Technology Perspectives*, October 19, 1998.

Prichett, J., *TRW Demonstrates World's Fastest Digital Chip; Indium Phosphide Technology Points To Higher Internet Speeds*, Hardware Telecommunications Internet Product Tradeshow, TRW. Inc., 2000.

Raghavan, G., M. Sokolick, W. E. Stanch, "Indium Phosphide ICs Unleash the High-Frequency Spectrum," *IEEE Spectrum*, October 2000.

Reed, M. A., and J. M. Tour, "Computing with Molecules," *Scientific American*, June 2000.

Rodwell, M., "Bipolar Technologies and Optoelectronics," 1999 IEEE MTT-S Symposium Workshop Technologies for the Next Millennium.

Science Wise, "Terahertz Quantum Well Emitters and Detectors," *Sciencewise.com*, April 14, 2001.

Stix, G., "The Triumph of the Light," *Scientific American*, January 2001.

Tuschman, R., "Bursting at the Seams," *IEEE Spectrum*, January 2001.

Zorpette, G., "The Quest for the Spin Transistor," *IEEE Spectrum*, December 2001.

Web resources

http://www.britannica.com/

The electromagnetic spectrum
http://imagine.gsfc.nasa.gov/docs/science/know_l1/emspectrum.html
http://observe.ivv.nasa.gov/nasa/education/reference/emspec/emspectrum.html

U.S. Frequency Allocation Chart
http://www.ntia.doc.gov/osmhome/allochrt.html

Optical Networking News
www.lightreading.com

2 FUNDAMENTALS OF ELECTRIC FIELDS

THE ELECTRIC FORCE FIELD

To understand high-frequency and RF electronics, you must first have a good grasp of the fundamentals of electromagnetic fields. This chapter discusses the electric field and is the starting place for understanding electromagnetics. Electric fields are created by charges; that is, charges are the source of electric fields. Charges come in two types, positive (+) and negative (–). Like charges repel each other and opposites attract. In other words, charges produce a force that either pushes or pulls other charges away. Neutral objects are not affected. The force between two charges is proportional to the product of the two charges, and is called Coulomb's law. Notice that the charges produce a force on each other without actually being in physical contact. It is a force that acts at a distance. To represent this "force at distance" that is created by charges, the concept of a *force field* is used. Figure 2.1 shows the electrical force fields that surround positive and negative charges.

By convention, the electric field is always drawn from positive to negative. It follows that the force lines emanate from a positive charge and converge to a negative charge. Furthermore, the electric field is a normalized force, a force per charge. The normalization allows the field values to be specified independent of a second charge. In other words, the value of an electric field at any point in space specifies the force that would be felt if a unit of charge were to be placed there. (A unit charge has a value of 1 in the chosen system of units.)

Electric field = Force field as "felt" by a unit charge

To calculate the force felt by a charge with value, q, we just multiply the electric field by the charge,

$$\vec{F} = q \cdot \vec{E}$$

Figure 2.1 Field lines surrounding a negative and a positive charge. Dotted lines show lines of equal voltage.

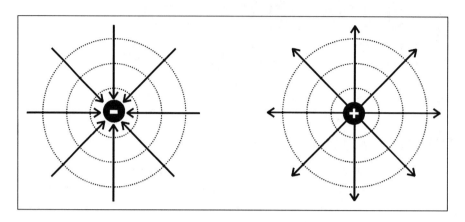

The magnitude of the electric field decreases as you move away from a charge, and increases as you get closer. To be specific, the magnitude of the electric field (and magnitude of the force) is proportional to the inverse of the distance squared. The electric field drops off rather quickly as the distance is increased. Mathematically this relation is expressed as

$$E = \frac{q}{r^2}$$

where r is the distance from the source and q is the value of the source charge. Putting our two equations together gives us Coulomb's law,

$$F = \frac{q_1 \cdot q_2}{r^2}$$

where q_1 and q_2 are the charge values and r is the distance that separates them. Electric fields are only one example of fields.

OTHER TYPES OF FIELDS

Gravity is another field. The gravitational force is proportional to the product of the masses of the two objects involved and is always attractive. (There is no such thing as negative mass.) The gravitational field is much weaker than the electric field, so the gravitational force is only felt when the mass of one or both of the objects is very large. Therefore,

Figure 2.2 Two balls attached by a spring. The spring exerts an attractive force when the balls are pulled apart.

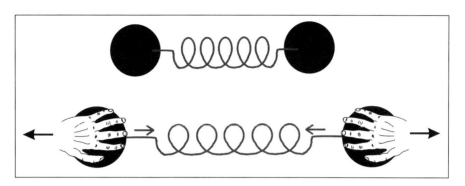

our attraction to the earth is big, while our attraction to other objects like furniture is exceedingly small.

Another example of a field is the stress field that occurs when elastic objects are stretched or compressed. For an example, refer to Figure 2.2. Two balls are connected by a spring. When the spring is stretched, it will exert an attractive force on the balls and try to pull them together. When the spring is compressed, it will exert a repulsive force on the balls and try to push them apart. Now imagine that you stretch the spring and then quickly release the two balls. An oscillating motion occurs. The balls move close together, then far apart and continue back and forth. The motion does not continue forever though, because of friction. Through each cycle of oscillation, the balls lose some energy until they eventually stop moving completely. The causes of fiction are the air surrounding the balls and the internal friction of the spring. The energy lost to friction becomes heat in the air and spring. Before Einstein and his theory of relativity, most scientists thought that the electric field operated in a similar manner. During the 1800s, scientists postulated that there was a substance, called *aether*, which filled all of space. This aether served the purpose of the spring in the previous example. Electric fields were thought to be stresses in the aether. This theory seemed reasonable because it predicted the propagation of electromagnetic waves. The waves were just stress waves in the aether, similar to mechanical waves in springs. But Einstein showed that there was no aether. Empty space is just that—empty.* Without any aether,

---

*This statement is not really true in quantum physics, which states that even the vacuum contains fluctuations of virtual particles. Refer to Chapter 6 for more information.

there is no way to measure absolute velocity. All movement is therefore relative.

VOLTAGE AND POTENTIAL ENERGY

A quantity that goes hand in hand with the electric field is *voltage*. Voltage is also called potential, which is an accurate description since voltage quantifies potential energy. Voltage, like the electric field, is normalized per unit charge.

Voltage = Potential energy of a unit charge

In other words, multiplying voltage by charge gives the potential energy of that charge, just as multiplying the electric field by charge gives the force felt by the charge. Mathematically we represent this by

$$U = q \cdot V$$

Potential energy is always a relative term; therefore voltage is always relative. Gravity provides a great visual analogy for potential. Let's define ground level as zero potential. A ball on the ground has zero potential, but a ball 6 feet in the air has a positive potential energy. If the ball were to be dropped from 6 feet, all of its potential energy will have been converted to kinetic energy (i.e., motion) just before it reaches the ground. Gravity provides a good analogy, but the electric field is more complicated because there are both positive and negative charges, whereas gravity has only positive mass. Furthermore, some particles and objects are electrically neutral, whereas all objects are affected by gravity. For instance, an unconnected wire is electrically neutral, therefore, it will not be subject to movement when placed in an electrical potential. (However, there are the secondary effects of electrostatic induction, which are described later in the chapter.)

Consider another example, a vacuum tube diode, as shown in Figure 2.3. Two metals plates are placed in an evacuated glass tube, and a potential (10 V) is placed across them. The negative electrode is heated. The extra electrons in the negative electrode that constitute the negative charge are attracted to the positive charge in the positive electrode. The force of the electric field pulls electrons from the negative electrode to the positive electrode. (The heating of the electrode serves to "boil off" electrons into the immediate vicinity of the metal.) Once free, the elec-

Figure 2.3 A vacuum tube diode, showing electrons leaving the negatively biased cathode to combine with positive charge at the anode.

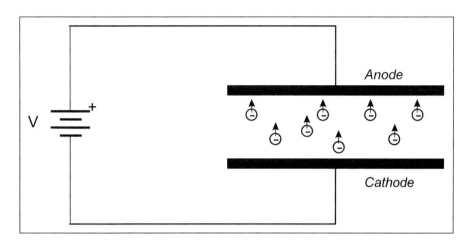

trons accelerate and then collide with the positive electrode where they are absorbed. Just before each electron collides, it is traveling very fast because of the energy gained from the electric field. Its kinetic energy can be easily calculated, in terms of electron-Volts (eV) and in terms of Joules (J):

$$U = q \cdot V = e \cdot (10\,\text{volts}) = 10\,\text{eV}$$
$$= 10 \cdot (1.6 \times 10^{-9}\,C) \cdot V = 1.6 \times 10^{-18}\,J$$

From this example, you can see how natural the unit of electron-Volts (eV) is for describing the energy of an electron.

The most important thing for you to remember is that voltage is a relative term. On a 9 volt battery, the (+) contact has a voltage of 9 volts relative to the (−) contact. Furthermore, the battery has a net charge of zero, although the charge is separated into negative and positive regions. The charge on the positive side is drawn to the charge on the negative side. Connecting a wire between the terminals allows the charges to recombine. The same result is true for a charged capacitor. What would happen if you brought a neutral unconnected wire close to one of the battery's terminals? Nothing, the wire is neutral. It has no net charge. We'll revisit the details of this situation a little later in the chapter.

CHARGES IN METALS

In electronics you will only encounter two types of charged particles, electrons and ions. To understand each, let's review the basic building blocks of matter. Matter consists of tiny particles called atoms. In each atom is a core or nucleus that contains protons and neutrons. The nucleus is very compact. Surrounding the nucleus are electrons. For a neutral atom, there are equal numbers of electrons and protons. The protons possess positive charge, and the electrons possess an equal but opposite charge. The neutrons in the nucleus are neutral. The electrons orbit the nucleus in a special way. You might imagine the electron as a small ball orbiting the nucleus in the same way that planets orbit the sun. However, this analogy is not quite correct. Each electron is smeared out in a three-dimensional cloud called an *orbital*. Atoms can lend out or borrow electrons, which leaves the atom with a net charge. Such atoms are called *ions* and they can be positively charged (missing electrons) or negatively charged (extra electrons). Ions of opposite charge can attract one another and form ionic bonds. These bonded ions are called *molecules*. Table salt, NaCl, is a good example of molecules. Each salt molecule consists of a positive ion (Na^+) and a negative ion (Cl^-). There also exist other types of molecular bonds. For instance, covalent bonds are established when two atoms share an electron.

Metal materials have special properties that make them good conductors of electricity. First of all, metals are crystals; that is, metals have an orderly construction of atoms. Most people who have not studied solid-state or semiconductor physics find it very surprising that metals are crystals, because we tend to associate crystals with transparent materials like quartz. As with all crystals, the structure of a metal is a three-dimensional lattice like that shown in Figure 2.4. Metals have a rather interesting bonding structure. The positive ions of the metal are held together by a sea of electrons that is shared by the entire crystal. Each atom of the metal typically contributes one or two electrons to this sea. You can picture a metal as closely packed balls (the ions) in a gas of small particles (the electrons). Because the "sea" electrons are free to roam throughout the metal, they serve to conduct electricity quite well. It then makes sense to call them conduction electrons to differentiate them from the electrons that are bound to the ions.

At the microscopic level, a metal has a lot going on, even without any applied electric field. Thermal vibrations cause the lattice of ions to vibrate and cause the conduction electrons to move around. With higher temperatures, the vibrations in the metal get larger. This effect stems

Figure 2.4 Example of atoms arranged in a crystal lattice structure (simple cubic).

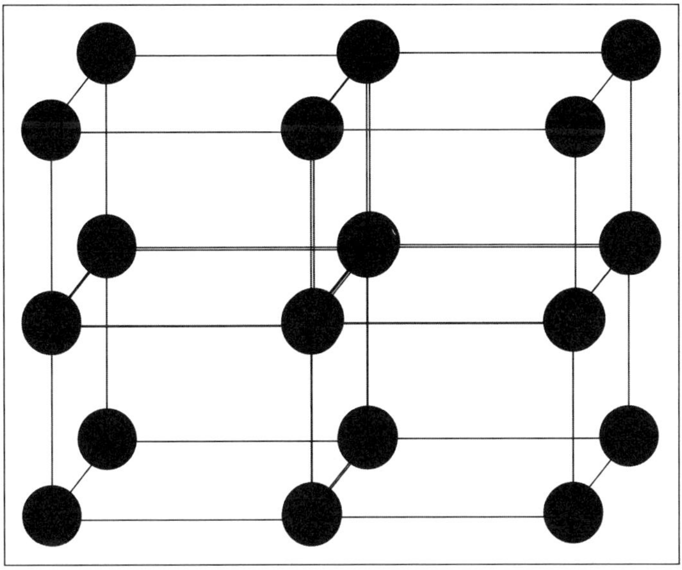

from the microscopic relationship of temperature. Temperature is proportional to the average kinetic energy of the particles,

$$T = \frac{m}{3k} v_{rms}^2$$

where k is Boltzman's constant, m is the atomic mass, and v_{rms}^2 is the root-mean-square particle velocity. (This formula applies directly to ideal gases, but it gives a decent approximation for the temperature of metals.) These thermal vibrations result in the random agitation of electrons and are sources of thermal noise in electronics. Every wire and resistor in a circuit injects white noise (i.e., with the same power level at all frequencies) from its internal thermal vibrations. Now if we place a voltage across a metal wire (or resistor), an electric field is developed through the wire. The electric field causes the conduction electrons to move more toward the positive (+) end of the wire. In other words, a current is produced. This current is a direct current (DC) so it has a constant value.

Current is defined as the amount of charge through a cross section of wire, per second. This definition implies that the charges are moving at a steady, constant value. However, this assumption is incorrect. Only the statistical average of the charge movement is a constant value. Each individual charge's movement is very random. The electrons are constantly colliding with the vibrating lattice ions and the other electrons. On average an electron in copper at room temperature only spends about 25 fsec between collisions (1 fsec equals 1/1,000,000,000,000,000 sec.). The electrons have a very large velocity between collisions, about 100,000 m/sec. However most (99.999999%) of the velocity is in a random direction. Statistically, the random components of the velocity cancel each other out, leaving a much smaller average velocity. Therefore, while the root-mean-square (RMS) velocity is very high, the average velocity is much lower. The average velocity of the electrons, which is in the direction of the applied field, is very small. This average velocity is called the drift velocity, and it accounts for the observed current. For a 1 meter long copper wire with 1/10 volt across it, the drift velocity is about 0.5 mm/sec or 5 feet per hour! It will therefore take an electron about 38 minutes to travel from one side of the wire to the other. Assuming this copper wire is 20 gauge wire, its resistance is 33 mΩ, and the current from 0.1 volts is 3 amps. Even though the electrons travel very slowly, their effect adds up to large currents because there are so many of them. In a 1 meter long 20 gauge copper wire, there are about 4.4×10^{22} conduction electrons. Now that is strength in numbers!

THE DEFINITION OF RESISTANCE

Besides the flow of DC current, something else happens when you apply a voltage across a conductor: it heats up. The temperature of the conductor rises because DC flow causes the electrons to have higher-energy collisions (on average) than they would if the voltage were not present. Since it requires energy to heat something up, energy must be provided by the voltage supply. The energy per second lost to heat is exactly equal to the power calculated from the power law of electric circuits:

$$I^2 \cdot R = \frac{V^2}{R}$$

In a rather roundabout way, we have arrived at the definition of *resistance*. Resistance quantifies the power that is lost to heating when a voltage or current is applied to the conductor.

ELECTRONS AND HOLES

Imagine that you have a solid metal ball or sphere and you place a charge on it. For a negative charge, this process equates to adding an excess of electrons to the sphere. For a positive charge, this equates to removing some electrons, leaving an excess of positive ions or *holes*, as they are called in semiconductor parlance. Holes can be thought of as virtual positive particles, which can move around in a material like a bubble moving in water. Whereas negative charges (electrons) move freely about the material, holes must move by means of charge theft. Let's say there is an atom that is missing one electron. This atom is therefore a positive ion (+), or equivalently, this atom is carrying a hole. This ion can "steal" a bound electron from a neighbor. (Remember that not all atoms are ionized at all times.) By stealing an electron from its neighbor, it has in effect given the hole to the neighbor. This way holes can move through the material, and can be thought of as positive particles in their own right.

In semiconductors like silicon, the holes and electrons can move with approximately equal ease. In other words, the hole and electron mobilities are approximately equal in silicon. In most metals, however, the electrons are very mobile, and the holes have negligible mobility. The holes are virtually stationary and only the negatively charged electrons can move. Negative charge in metals is created when electrons move away from a region, leaving positively charged holes behind. Positive charge in metals should be thought of as a lack of free electrons. Most often, when we are not so formal, we do talk about positive charges moving about in a metal; just keep in mind that in most metals, it is always the negatively charged electrons that do the moving.

Back to the problem of the metal ball. When charge is placed onto the ball, the individual charges will immediately spread apart as far as possible because like charges repel each other. The upshot is that all the charge becomes concentrated at the surface. It's like a bunch of people in a large room. If each person tries to avoid the rest, they will migrate to the walls of the room, like a "wallflower" at a high school dance. How quickly charges distribute themselves to the surface is proportional to the *relaxation time* of the material. In the present context, the relaxation time can be approximated as the dielectric constant (discussed later in the chapter) of the material divided by the conductivity of the material. The relaxation time specifies how freely charge can move in a material. For copper, the relaxation constant is

$$\tau = \frac{\varepsilon}{\sigma} = \frac{8.85 \times 10^{-12}\, \frac{F}{m}}{5.8 \times 10^{7}\, \frac{1}{\Omega \cdot m}} = 1.5 \times 10^{-19}\, \text{sec}$$

Therefore, charge placed on a copper object will very quickly redistribute so that it all resides on the surface. The charge half-life of a material is about 0.7 times the relaxation time. An example will help illustrate this concept. If charge is somehow placed at the center of a metal ball, the charge will immediately start to migrate toward the surface. After a half-life in time (0.7τ), half of the charge will have migrated away from the center.

ELECTROSTATIC INDUCTION AND CAPACITANCE

To understand capacitance, you need to first understand the process of *electrostatic induction*. For example, consider that you have a metal ball that is positively charged, near which you bring a neutral metal ball. Even though the second ball has overall neutrality, it still contains many charges. Neutrality arises because the positive and negative charges exist in equal quantities. When placed next to the first ball, the second ball is affected by the electric field of the charged ball. The charges of the second ball separate. Negative charges are attracted, and positive charges are repelled, leaving the second ball polarized, as shown in Figure 2.5. This polarization of charge is called electrostatic induction.

A direct consequence of electrostatic induction is that the electric field inside an unconnected conductor is always zero at steady state. When a conductor is first placed in a field, the field permeates the conductor. The charges then separate as described in the preceding paragraph. The separation of charge tends to neutralize the electric field. Charge movement continues until the electric field reaches zero. Another way of stating this is that the voltage inside a conductor is constant at steady state. Placed in an ambient electric field, the conductor quickly adjusts its charge configuration until it has reached the voltage potential of its environment.

Now let's connect a metal object to the second ball using a wire. As shown in Figure 2.6, the charge polarizes even further, with the negative charge of the neutral objects moving as far away as possible. Figure 2.7 takes this one step further by connecting the earth to the second ball. (An earth connection can be achieved by connecting the ball to the third prong of a wall outlet, which typically is "earthed" on the

Figure 2.5 A) A negatively charged metal sphere. B) A neutral sphere (right) is brought close to the negative sphere (left).

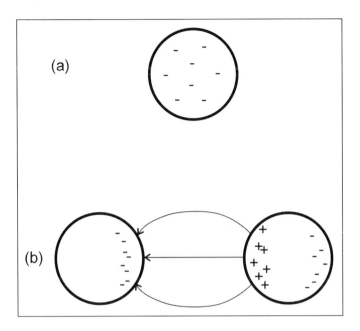

Figure 2.6 A second neutral object is connected by a wire.

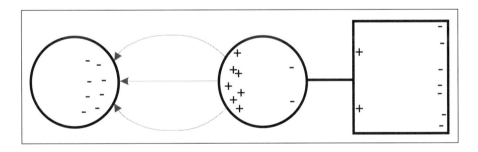

outside of each building using 8 foot or longer copper stakes.) Here the negative charge will move down the wire, into the earth, and go very far away.

Instead of placing a constant charge on the first metal ball, you could connect an oscillating charge, as in Figure 2.8. For example, assume that the AC voltage is at a frequency of 60 Hz; the polarization induced

Figure 2.7 The neutral sphere is connected to ground with a wire.

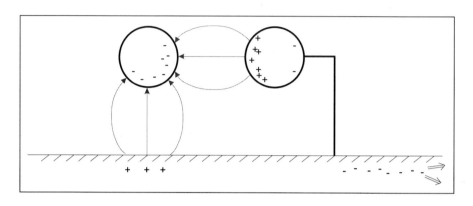

Figure 2.8 An AC source is connected to the first sphere.

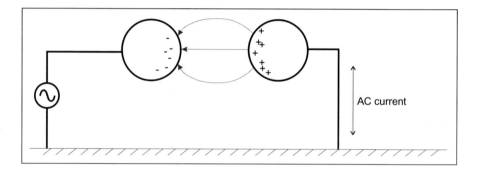

in the second ball will alternate at a frequency of 60 Hz. The alternating polarization will also cause current to flow in the wire that connects the second ball to the ground. What you have created is simply a *capacitor!* Any metal conductors that are separated by an insulator form a capacitor. In Figure 2.9, the two balls have been replaced with metal plates, forming a more familiar and efficient capacitor. Notice that no current actually traverses the gap between the plates, but equal current flows on both sides, as charge rushes to and from the plates of the capacitor. The virtual current that passes between the plates is called *displacement current*. It is really just a changing electric field, but we call it a current.

You can get a good feel for electrostatic induction by learning how some simple, but ingenious, inventions work. If you have ever worked

Figure 2.9 Spheres are replaced by metal plates.

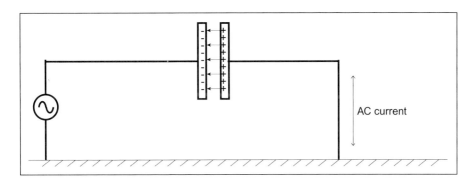

Figure 2.10 Block diagram for a non-contact field detector.

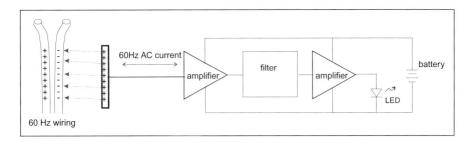

on the electric *mains* wiring that runs in the walls of your house and provides the 120 V power to your electric outlets and lamps, you have probably purchased what is called a *non-contact field detector*. (Incidentally, the term "mains" refers to the fact that your power is provided by your electric utility company. The same term is also used for the water that enters your house from the water utility.) A non-contact field detector is a device that is about the size and shape of a magic marker. It allows you to determine if a wire is live (is connected to voltage) or not without actually making metallic contact. This nifty invention can be waved near insulated wires, or it can be inserted into an electric outlet. It detects live wires and live outlets using the phenomenon of electrostatic induction that you just learned about. A simple schematic of such a device is shown in Figure 2.10. A metal lead or plate is attached to the input of a high-impedance amplifier. This plate serves one side of a capacitor. When the plate is brought close to an object that has an electric field, charge is induced on the plate. Some of the charge has to pass through the amplifier. Because the amplifier has high input impedance, a reasonable voltage will be created at its output. If the object being tested has a varying electric field, like a wall outlet, an AC current will be induced in the device. The rest of the circuitry serves to light an LED when the induced current varies continuously at around 60 Hz.

A variation of this circuit uses a human as the capacitive plate of the device. Glow tube meters and zero-pressure "touch" buttons work in this manner. A glow tube meter is a device for testing electrical outlets. It looks like a screwdriver, except the handle is made of clear plastic and contains a small glow tube inside. The glow tube is a glass tube filled with neon gas and contains two separated electrodes. One electrode is connected to the screwdriver blade and makes contact inside the electric outlet. The other electrode is connected to a piece of metal at the end of the screwdriver handle. By touching your finger to the handle end, you become part of a circuit. Being a decent conductor, your body serves the function of being one half of a capacitor. A small amount of the 60 Hz displacement current (~10 µA) is able to capacitively couple from your body to the ground wires in the wall. This small current causes a voltage drop across you and across the glow tube. The gas inside the glow tube ionizes and conducts electricity. In the process, it gives off a faint orange glow, telling you that the outlet is working. You certainly should use care with such a device, and be sure not to touch the screwdriver shaft directly.

Touch buttons are found in some elevators. These buttons are metal, and if you encounter one, you will notice that the button does not physically depress when you touch it. This button is connected to a high-impedance amplifier. When you touch the button, you again form a capacitor plate. In this application, your body couples 60 Hz energy from the wires in the elevator and the fluorescent lights on the ceiling to the metal touch pad. A high-impedance amplifier amplifies the current and determines that you have, for instance, touched the button for the third floor.

It may be tempting to think of these applications as antennas picking up 60 Hz radiation, but this idea is incorrect. Being electrically small wires, virtually no energy is radiated. In addition, you are standing well within the near field of the source. These topics are discussed in detail in Chapter 5.

INSULATORS (DIELECTRICS)

To make an electronic circuit work, you not only need conductors, you also need insulators. Otherwise, every part of the circuit would be shorted together! A perfect insulator is a material that has no free charge. Therefore, if a voltage is placed across an insulator, no current will flow. Even though no current flows, that does not mean that the dielectric does not react to electric fields. Most insulators, also known as dielectrics, become internally polarized when placed in an electric field. The internal polarization occurs from rotation of molecules (in liquids and gases) or from distortion of the electron cloud around the atoms (in solids).

A simple example is the polarization of H_2O molecules in liquid water. An H_2O molecule is shown in Figure 2.11. Due to the structure of the molecule, the charge is not exactly symmetric. Without an external electric field applied, the molecules have random orientations (polarizations) due to thermal motion. When a field is applied, the molecules tend to line up so that their negative sides are facing the positive voltage. The molecules therefore set up a secondary electric field that opposes the direction of the applied field. The result is that inside the dielectric material, the net electric field is reduced in value. The ratio of the applied field to the reduced field is called the (relative) dielectric constant, ε_r. This value is relative to the permittivity of a vacuum, $\varepsilon_o = 8.85 \times 10^{-12}$ F/m. The dielectric constant of a material has a direct effect on capacitance. For example, a capacitor made from two parallel plates has a capacitance,

$$C = \frac{\varepsilon_r \varepsilon_o A}{d}$$

where A is the area of each plate, d is the distance between the plates, and ε_r is the dielectric constant of the material sandwiched between the plates. As you can see from this formula, increasing the dielectric constant causes an increase in capacitance. Therefore, if you want to make a capacitor that is physically small but has a large capacitance, use a high dielectric constant material between the plates. It follows that a higher dielectric constant corresponds to higher energy storage.

Although a perfect dielectric has no free charge, even the best insulators in the real world will have some free charge. Even air conducts electricity, albeit poorly.

STATIC ELECTRICITY AND LIGHTNING

Let's go back to the charged sphere of Figure 2.5. If the second metal ball is placed closer to the charged ball so that they actually touch, the

Figure 2.11 A) Water molecule. B) Water molecules in random orientation with no field applied. C) Water molecules line up when an electric field is applied.

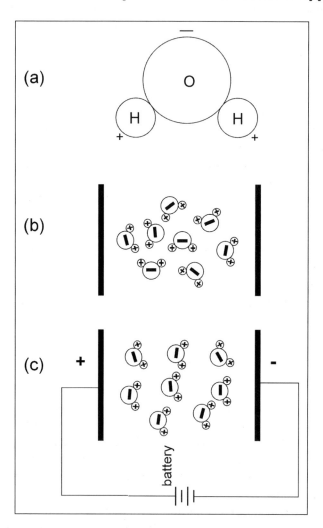

negative charge will now redistribute itself evenly over the two balls. If the balls are moved apart again, each will now have half of the original negative charge. If balls of unequal size were used, then more charge would end up on the larger ball, because it has a larger surface area. If the charge is large enough, very high voltages will be produced as the balls are brought close together. When the electric field exceeds the breakdown strength of air (typically about 5000 volts per centimeter), a spark will occur between the two balls, allowing charge to transfer without physical contact. In this phenomenon, air molecules become ionized, forming a jagged conducting path between the two conductors. In the process, some of the original charge is lost to the air ions. What you have just learned is the process of *static discharge* (illustrated in Figure 2.12). The same effect occurs when you walk across the carpet and then get shocked by a doorknob. Your body becomes charged while walking across the carpet. Called the *triboelectric* effect, it consists of charge separation when certain materials are placed and/or rubbed together and then pulled apart. The outcome is that your body takes up a charge and the carpet stores an equal but opposite charge. Typical static discharges you encounter will be in the range of 5 kV to 15 kV and will produce a peak current of about 1 amp! Quite a large jolt.

You may wonder why such high voltage and current hurts but is not dangerous. There are several reasons for this. First of all, the entire discharge only lasts about 1 μsec. For this same reason, shocks from spark plugs *usually* cause no harm. Second, the current is mostly concentrated near the discharge (i.e., your finger) and then quickly spreads and dissipates. Negligible current will traverse your heart, which is the danger zone for current in the human body. Typically, a person can sense currents of 1 mA or more, with currents over 40 mA being possibly lethal as they pass through the heart. Therefore, it is how much current flows through the body and where it flows that is of concern. When a person becomes part of a circuit, there are four parameters that determine the current: 1) source voltage, 2) source resistance, 3) contact resistance, and 4) internal body resistance for the current path. The second two parameters are a function of body physiology. The contact resistance is mainly caused by the dead skin layer where contact is made. Contact resistance is typically on the order of 100 ohms for sweaty skin to 100 kohms for very dry skin. Beware that if the skin is cut, the contact resistance becomes negligible. The internal body resistance is fairly low due to the fact that nerves and blood vessels make good conductors. Any limb-to-limb internal resistance can be approximated as about 500 ohms. The effects of electricity are felt differently, depending on frequency. The most dangerous frequency range is from about 5 Hz to about 500 Hz and peaks in danger right about 60 Hz, the frequency of power lines. The frequency sensitivity has to do with the physiology of the human nervous system, which typically communicates via pulse trains in this range of frequencies. A 60 Hz current has approximately two to three times the danger as the same current at DC. As opposed to DC signals, these AC signals can cause muscles to lock up, leaving a person unable to let go of the voltage source.

Figure 2.12 A) A neutral sphere (right) is brought close to the negative sphere (left). B) The spheres become very close and a path of ionized molecules forms in the air, causing a spark. C) After the spark has dissipated, some of the charge has been carried away by the ions in the air. The remaining charge is split equally between the spheres.

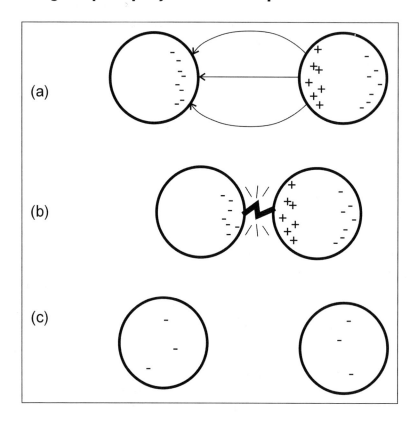

Lightning is static electricity on a grand scale. In the large circulating winds of a thunderstorm, raindrops and ice crystals collide, causing charge separation. The process of charge separation is not fully understood, but it is known that positive charge collects at the top of the cloud and negative charge collects at the bottom of the cloud. The negative charge at the bottom of the cloud induces an equal but opposite region of charge on the ground below. You can think of it as a localized charge shadow. When the charge builds up to a high enough voltage (typically 10 MV to 100 MV), the air starts to ionize in the form of a jagged "leader" which migrates from the cloud to the ground in discrete jumps of (typically) a few hundred feet. When the ionized leader gets one jump away from the ground, its path is affected by the objects within the immediate vicinity, often connecting to the highest projecting object. Upon contact, a conducting path then connects the cloud with the ground, allowing the cloud to discharge. Peak currents can range from a few kAmps to 150 kAmps, and the event typically lasts about several hundred microseconds. With such high currents, lightning is always dangerous to people.

So what happens to all this negative charge that accumulates on the ground? It gradually migrates back to the atmosphere through the small concentration of ions always present in air. During fair weather, the upper atmosphere has a positive charge and the earth has a negative charge, forming a giant capacitor (about 5000 Farad) that is discharging an average of 1800 amps at any given time. This "fair weather current" is needed to balance out the currents from thunderstorms. We are standing in the middle of it! Consequently, the air of a typical day has a DC electric field in it of about 100 volts/meter. Even though we are in the middle of this high field, we don't experience much of a problem because, being very good conductors when compared to air, we locally short out the field. Approximating head-to-toe human body internal resistance as 700 ohms and the air resistance of a 6 foot long, 3 foot by 3 foot columnar region as about 10^{14} ohms, an equivalent circuit can be constructed. We end up being a very small resistance in series with a very large resistance (the miles of air between a person and upper atmosphere). Using these assumptions, the voltage across the body is approximated as $3\mu V$.

Large electric fields can cause other interesting effects. As mentioned earlier, large electric fields can cause gases to ionize; that is, electrons are freed from the gas atoms, leaving charged ions behind. Ionized gases conduct electricity and produce visible light in the process. Electrical *corona* is the term used to describe the glowing region of ionized gas that can occur around conductors. The glow is visible radiation produced when an atom gives up or accepts an electron. Coronas are created near conductors that have high electric fields emanating from them. High-voltage power lines must be placed far enough apart to avoid creating coronas since a corona consumes energy. Fluorescent light bulbs work via the same phenomena. A high voltage causes the gas inside the bulb to ionize. In fact, an unconnected fluorescent light bulb will glow if held in air near high-voltage wires! The large electric field causes the gas inside the bulb to ionize, which in turn causes the visible light.*

---

*To be exact, the gas emits UV radiation, which is absorbed and then reradiated by the fluorescent powder on the inside surface of the bulb. The reradiated light is in the visible region of the spectrum.

Figure 2.13 Electric field surrounding a capacitor with DC voltage applied. This figure was created using Ansoft Corporation's Maxwell 2D field solver software (http://www.ansoft.com).

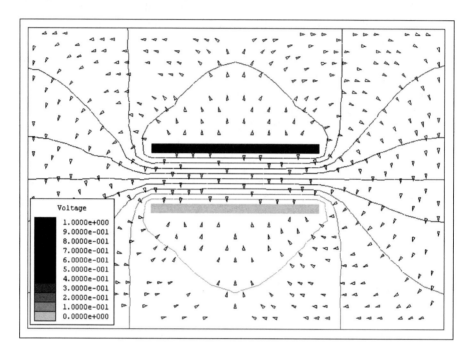

Pointed objects such as flagpoles and ship masts tend to concentrate electric charge and can produce a corona during a strong thunderstorm. Before this effect was understood, sailors thought that it was the sign of a ghost or spirit and called it *St. Elmo's fire*. If you see a corona during a storm, this is a bad sign, since whatever object is producing the corona is acting as a lightning rod and will attract a lightning bolt if one approaches the area. A *lightning rod* is a metal rod that is connected to earth and protects a house or structure by attracting nearby lightning strikes. The lightning rod conducts the lightning current safely into the earth, preventing it from finding alternative paths to ground (like through your roof). It is a common myth believed by many engineers and scientists that a lightning rod prevents lightning by slowly discharging the cloud immediately above. Let me emphasize that this myth has been proven false many times over by lightning researchers. In fact, it would take over 800 hours for a lightning rod to discharge the typical

Figure 2.14 Electric field surrounding another capacitor with DC voltage applied. This figure was created using Ansoft Corporation's Maxwell 2D field solver software (http://www.ansoft.com).

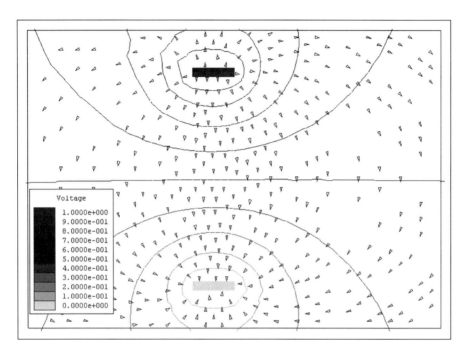

cloud, and most of the ions the lightning rod releases get dispersed by the high storm winds.

The triboelectric effect can cause other, less dramatic problems. Moving cables can cause noise via the triboelectric effect. The movement of the cable causes friction between the insulation and the metal, rubbing charge from the insulation. A similar effect can happen with outdoor cables or antennas that are blown by the wind, especially if the weather is stormy and the air is well ionized.

THE BATTERY REVISITED

Earlier in this chapter, during the introduction of voltage, I mentioned that a neutral conductor like a wire or metal plate that is brought close to a battery will not be affected by the battery. Even if the battery or

Figure 2.15 Electric field of DC voltage across a conducting wire. Arrowheads show the direction of the electric field. Voltage is shown by shade inside the conductor and as contours outside the conductor. This figure was created using Ansoft Corporation's Maxwell 2D field solver software (http://www.ansoft.com).

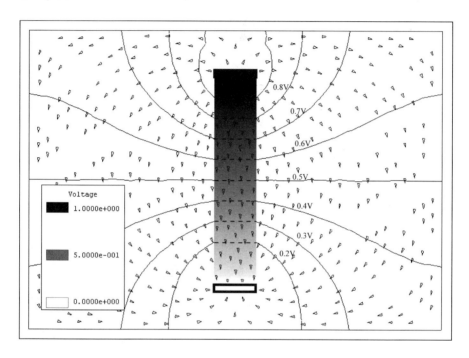

capacitor has a very high voltage, you will not see a spark (static discharge) develop to a third conductor. Why not?

To avoid discussing the details of the electrochemistry inside a battery, let's assume that we have a capacitor charged to a large voltage. One plate has many positive charges and the other plate has many negative charges. The net charge of the capacitor is neutral. Now bring a neutral, isolated metal ball near to the negative terminal. A slight charge is induced in the ball, but not much because most of the negative charge has a greater attraction to the other plate of the capacitor. Furthermore, we will not see a spark because the capacitor as a whole has a neutral charge. Suppose that charge did start to leave the capacitor to jump to the neutral ball. The capacitor would then have a charge imbalance and the charge would be attracted right back.

Table 2.1 Low-Frequency (DC) Conductivity and Permittivity of Various Materials

Material	Conductivity, $1/[\Omega m]$	Permittivity (Dielectric Constant $\times \varepsilon_0$)
Copper	5.7×10^7	$1 \times \varepsilon_0$
Stainless steel	10^6	$1 \times \varepsilon_0$
Salt water	~4	$80 \times \varepsilon_0$
Fresh water	~10^{-2}	$80 \times \varepsilon_0$
Distilled water	~10^{-4}	$80 \times \varepsilon_0$
Animal muscle	0.35	[no data]
Animal fat	4×10^{-2}	[no data]
Typical ground (soil)	~10^{-2} to 10^{-4}	$3 \times \varepsilon_0$ to $14 \times \varepsilon_0$
Glass	~10^{-12}	~$5 \times \varepsilon_0$

$\varepsilon_0 = 8.85 \times 10^{-12}$ F/m

Table 2.1 data adapted from Krauss and Fleisch, *Electromagnetics with Applications*, 5th Edition, Boston: McGraw-Hill, 1999, and from Paul and Nasar, *Introduction to Electromagnetic Fields*, 2nd Edition, New York: McGraw-Hill, 1987.

ELECTRIC FIELD EXAMPLES

Figures 2.13 through 2.15 show examples of the electric fields surrounding various circuit configurations. Figure 2.13 shows the electric field surrounding a simple plate capacitor. Figure 2.14 shows the electric field of a plate capacitor whose plates are less wide and farther apart. Figure 2.15 shows the same structure, with a conductor inserted between the plates. This figure shows how the electric field looks inside a current-carrying wire.

CONDUCTIVITY AND PERMITTIVITY OF COMMON MATERIALS

To conclude this chapter on electric fields, I have listed the conductivity and dielectric constants of several common materials in Table 2.1. Note that salt water and body tissue are much more conductive than distilled water. The conductivity is caused by dissolved chemicals like salt, which dissolves into Na^+ and Cl^- when placed in water. These ions act to conduct electricity. Also notice that the typical ground is not a very good conductor.

Keep in mind that these values are for DC and low frequency. All materials, conductors and insulators, change characteristics at different frequencies. Most materials used in electronics have constant properties into the microwave region, and some materials such as metals, glass, and teflon don't change until the infrared. Materials such as water and soil change dramatically in the radio wave frequencies. You'll learn more about the frequency dependence of materials in Chapter 15.

BIBLIOGRAPHY: ELECTRIC FIELDS AND CONDUCTION

Cogdell, J. R., *Foundations of Electrical Engineering*, 2nd Edition, Englewood Cliffs, NJ: Prentice-Hall, 1995.

Eisberg, R., and R. Resnick, *Quantum Physics of Atoms, Molecules, Solids, Nuclei, and Particles*, 2nd Edition, New York: John Wiley & Sons, 1985.

Epstein, L. C., *Thinking Physics—Is Gedanken Physics; Practical Lessons in Critical Thinking*, 2nd Edition, San Francisco: Insight Press, 1989.

Feynman, R. P., R. B. Leighton, and M. Sands, *The Feynman Lectures on Physics Vol I: Mainly Mechanics, Radiation, and Heat*, Reading, Mass.: Addison-Wesley Publishing, 1963.

Feynman, R. P., R. B. Leighton, and M. Sands, *The Feynman Lectures on Physics Vol II: Mainly Electromagnetism and Matter*, Reading, Mass.: Addison-Wesley Publishing, 1964.

Glover, J. D., and M. Sarma, *Power System Analysis and Design with Personal Computer Applications*, Boston: PWS Publishers, 1987.

Griffiths, D. J., *Introduction to Electrodynamics*, 3rd Edition, Upper Saddle River, NJ: Prentice Hall, 1999.

Halliday, D., R. Resnick, and J. Walker, *Fundamentals of Physics*, 6th Edition, New York: John Wiley & Sons, 2000.

Heald, M., and J. Marion, *Classical Electromagnetic Radiation*, 3rd Edition, Fort Worth, Texas: Saunders College Publishing, 1980.

Jackson, J. D., *Classical Electrodynamics*, 2nd Edition, New York: John Wiley & Sons, 1975.

Kraus, J. D., and D. A. Fleisch, *Electromagnetics with Applications*, 5th Edition, Boston: McGraw-Hill, 1999.

Pierret, R. F., *Semiconductor Device Fundamentals*, Reading, Mass.: Addison-Wesley, 1996.

Ramo, S., J. R. Whinnery, and T. Van Duzer, *Fields and Waves in Communication Electronics*, 2nd Edition, New York: John Wiley, 1989.

Shadowitz, A., *The Electromagnetic Field*, New York: Dover Publications, 1975.

Ulaby, F. T., *Fundamentals of Applied Electromagnetics*, Englewood Cliffs, NJ: Prentice-Hall, 1999.

Vanderlinde, J., *Classical Electromagnetic Theory*, New York: John Wiley & Sons, 1993.

BIBLIOGRAPHY: STATIC ELECTRICITY AND LIGHTNING

Adams, J. M., *Electrical Safety a Guide to the Causes and Prevention of Electrical Hazards*, London: The Institution of Electrical Engineers, 1994.

Anderson, K., *Frequently Asked Questions (FAQ) About Lighting*, www.nofc.foresty.ca Edmonton: Canadian Forest Service.

Carlson, S., "Detecting the Earth's Electricity," *Scientific American*, July 1999.

Carlson, S., "Counting Atmospheric Ions," *Scientific American*, September 1999.

Carpenter, R. B., Jr., *Lightning Protection Requirements for Communications Facilities*, Lightning Eliminators & Consultants, Inc., Report No. T9408, August 1994.

Carpenter, R. B., Jr., and Y. Tu, *The Secondary Effects of Lightning Activity*, Lightning Eliminators & Consultants, Inc., January 1997.

Chalmers, J. A., *Atmospheric Electricity*, Oxford: Pergamon Press, 1967.

Diels, J.-C., R. Bernstien, K. E. Stahlkopf, and X. M. Zhao, "Lightning Control with Lasers," *Scientific American*, August 1997.

Encyclopedia Britannica Inc., "Electricity," *Encyclopedia Britannica*, Chicago: Encyclopedia Britannica Inc., 1999.

Feynman, R. P., R. B. Leighton, and M. Sands, *The Feynman Lectures on Physics Vol II: Mainly Electromagnetism and Matter*, Reading, Mass.: Addison-Wesley Publishing, 1964.

Frydenlund, M. M., *Lightning Protection for People and Property*, New York: Van Nostrand Reinhold, 1993.

Golde, R. H., *Lightning Protection*, New York: Chemical Publishing, 1973.

Jonassen, N., *Electrostatics*, New York: Chapman and Hall, 1998.

Kithil, R., *Lightning Rod Behavior: A Brief History*, National Lightning Safety Institute Facilities Protection, September 18, 2000.

Kraus, J. D., and D. A. Fleisch, *Electromagnetics with Applications*, 5th Edition, Boston: McGraw-Hill, 1999.

Moore, C. B., "Measurements of Lightning Rod Responses to Nearby Strikes," *Geophysical Research Letters*, Vol. 27, No. 10, May 15, 2000.

Sorwar, M. G., and I. G. Gosling, "Lightning Radiated Electric Fields and Their Contribution to Induced Voltages," *IEEE*, 1999.

Uman, M. A., *The Lightning Discharge*, Orlando, Florida: Academic Press, 1987.

Uman, M. A., and E. P. Krider, "Natural and Artificially Initiated Lightning," *Science*, Vol. 246, October 27, 1989.

Williams, E. R., "The Electrification of Thunderstorms," *Scientific American*, November 1988.

Web Resources

Animations of how fields interact with water molecules can be found at http://www.Colorado.EDU/physics/2000/microwaves/index.html

EMF and Human Health
http://www.dnai.com/~emf/

3 FUNDAMENTALS OF MAGNETIC FIELDS

The other half of electromagnetics is of course the magnetic field. Magnetic fields are inherently different and more difficult to grasp than electric fields. Whereas electric fields emanate directly from individual charges, magnetic fields arise in a subtle manner because there are no magnetic charges. Moreover, because there are no magnetic charges, magnetic field lines can never have a beginning or an end. Magnetic field lines always form closed loops.

You may have heard that some particle physicists have been searching for magnetic charges (or "magnetic monopoles," as the particle physicists call them) in high-energy experiments. In fact, many unified theories of physics require such particles. However, at this point such particles have not been found. Even if they were to be found, they would be so rare as to be inconsequential to everyone except the particle physicists and cosmologists. Instead of hoping for magnetic charges to bail us out, you need to just accept the fact that magnetic fields are inherently different from electric fields. In Chapter 6, when you learn about relativity, you will learn how the magnetic and electric field phenomena are related in the same manner as space and time. Just as space and time are very different but taken together form an interwoven entity, so do the electric and magnetic fields.

MOVING CHARGES: SOURCE OF ALL MAGNETIC FIELDS

Without magnetic charges, magnetic fields can only arise indirectly. In fact, all magnetic fields are generated indirectly by moving electric charges. It is a fundamental fact of nature that moving electrons, as well as any other charges, produce a magnetic field when in motion. Electrical currents in wires also produce magnetic fields because a current is basically the collective movement of a large number of electrons. A steady (DC) current through a wire produces a magnetic field that encircles the wire, as shown in Figure 3.1.

Figure 3.1 Magnetic field lines surrounding a current-carrying wire.

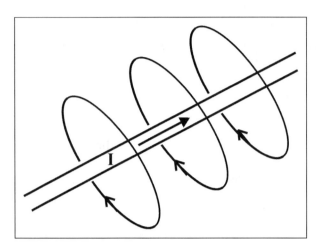

Figure 3.2 Magnetic field lines surround a moving electron.

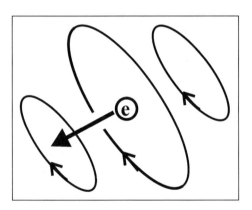

A single charge moving at constant velocity also produces a tubular magnetic field that encircles the charge, as shown in Figure 3.2. However, the field of a single charge decays along the axis of propagation, with the maximum field occurring in the neighborhood of the charge. The law that describes the field is called the Biot-Savart law, named after the two French scientists who discovered it.

It is interesting to note that if you were to move along at the same velocity as the charge, the magnetic field would disappear. In that frame

MAGNETIC DIPOLES

Figure 3.3 The right hand rule: the magnetic field (B) curls like the fingers of the hand around the current (I), which points in the direction of the thumb.

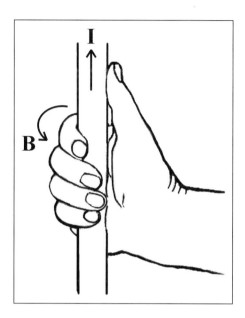

of reference, the charge is stationary, producing only an electric field. Therefore, the magnetic field is a relative quantity. This odd situation hints at the deep relationship between Einstein's relativity and electromagnetics, which you will learn about in detail in Chapter 6.

The magnetic field direction, clockwise or counterclockwise, depends on which direction the current flows. You can use the "right hand rule" for determining the magnetic field direction. Using Figure 3.3 as a guide, extend your hand flat and point your thumb in the direction of the current (i.e., current is defined as the flow of positive charge, which is opposite to the flow of electrons). Now curl the rest of your fingers to form a semicircle. The magnetic field will follow your fingers, flowing from your hand to your fingertips, or in other words, the arrow tips of the field will be at your fingertips.

MAGNETIC DIPOLES

Now, consider a current that travels in a loop, as shown in Figure 3.4. The magnetic field is a toroidal (donut-shaped) form. The magnetic field

Figure 3.4 Magnetic field lines surrounding a current loop.

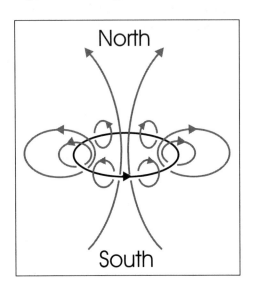

of this device flows out of one side and back in the other side. Although the field lines still form closed loops, they now have a sense of direction. The side where the field lines emanate is called the north pole, and the side they enter is called the south pole. Hence, such a structure is called a magnetic dipole. Now if a wire is wound in many spiraling loops, a solenoid like that shown in Figure 3.5 is formed. A solenoid concentrates the field into even more of a dipole structure.

Another example of a dipole is the simple bar magnet. The field of such a permanent magnetic is shown in Figure 3.6. This field is just like that of a solenoid, implying that there must be a net circular current inside the magnetic material. However, in this case the current is due to electron spin.*

The definitions of north pole and south pole come from the natural magnetic field that the earth produces. A sensitive magnetic dipole like a compass needle will rotate itself such that its north pole points towards the Earth's geographic north pole. The Earth's north pole is the side where the global magnetic field enters. The Earth's south pole is therefore the side from which the magnetic field emanates. (The Earth's

---

*Spin is an intrinsic quantum property of electrons. Spin describes the angular momentum of an electron. Logically, it follows that if an electron is spinning, then its charge will create a magnetic field like that of the earth.

Figure 3.5 Magnetic field lines surrounding a solenoid that is carrying a DC current.

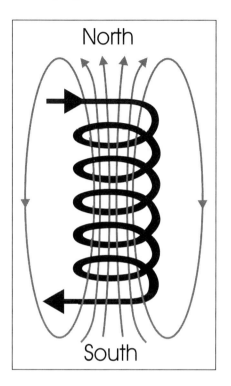

magnetic poles are therefore opposite to the geographic poles. The geographic north pole is the magnetic south pole and vice versa.) That's right, you guessed it, the Earth's magnetic field (shown in Figure 3.7) also arises from currents. In the case of the Earth, the currents are from charges revolving inside the Earth's molten core.

Even the electron has an inherent dipole magnetic field. An electron has an inherent angular momentum (called *spin*) and it certainly has charge. Although we don't know what an electron is or what really happens inside an electron, we can think of an electron as a spinning ball of charge that creates its own magnetic dipole, just like the rotating currents inside the Earth create its magnetic field. The magnetic dipole of an electron is quite small and we typically can ignore it when we study the movement of a free electron. However, the electron's magnetic field does play an important role when the electron is bound in the atomic structure of materials.

Figure 3.6   Magnetic field lines surrounding a bar magnet.

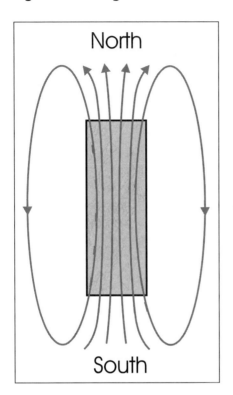

EFFECTS OF THE MAGNETIC FIELD

The Dipole

Now that you understand how magnetic fields are created, you need to understand how magnetic objects are affected by an external magnetic field. The situation is more complex than the electric field, where charges just follow the electric field lines. The effect of the magnetic field is rotational. To analyze how the magnetic field operates, you need some form of fundamental test particle. For the electric field, we use a point charge (i.e., a charged, infinitesimally small particle). Since magnetic charges do not exist, some alternative must be used. One such test particle is an infinitesimally small magnetic dipole. A magnetic dipole test particle can be thought of as a compass needle made exceedingly small.

A magnetic dipole has a north pole and a south pole, implying that it has direction in addition to magnitude. In other words, it is a vector

EFFECTS OF THE MAGNETIC FIELD

Figure 3.7 Magnetic field lines surrounding the earth.

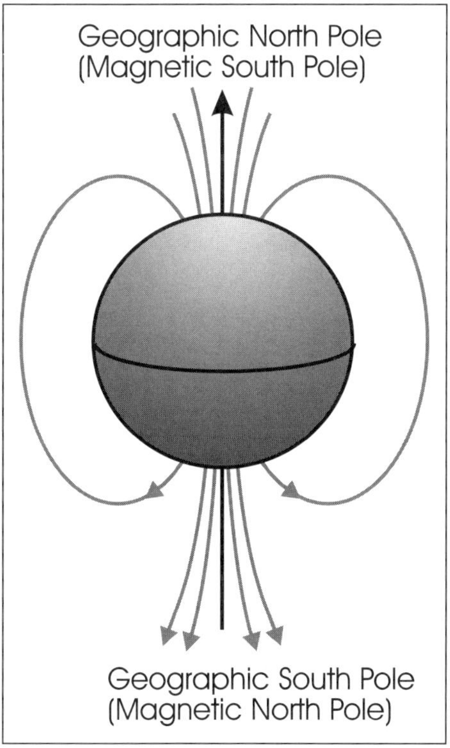

```
Geographic North Pole
(Magnetic South Pole)


Geographic South Pole
(Magnetic North Pole)
```

quantity. The property of direction highlights a fundamental characteristic of the magnetic field that makes it different from the electric field. You know from experience that a compass needle always rotates so that the marked end (north pole) of the needle points north. If we place our conceptual compass in a magnetic field, the needle will likewise rotate until it points along the field lines. Its orientation will be such that its field lines up with those lines of the field in which it is immersed. So instead of a force being transmitted to the test dipole, torque is transmitted. A torque is the rotational analogy to a force. In this instance, the magnetic field acts as a "torque field" in comparison to the electric force field. This relation can be mathematically expressed as the following cross product:

$$\vec{\tau} = \vec{\mu} \times \vec{B},$$

where τ is the torque in Newton-meters, μ is the magnetic dipole moment in ampere-meters^2, and B is the magnetic field in Webers/

Figure 3.8 The cross product right hand rule.

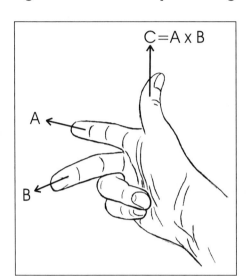

meter[2]. All three variables are vector quantities; that is, each has a magnitude and direction. The direction of the torque can be determined by the right hand rule for cross products, as shown in Figure 3.8. The magnitude of the torque is

$$\tau = \mu B \sin(\theta),$$

where θ is the angle between the dipole, μ, and the magnetic field, B.

Motors

The electric motor is the most common method for converting electromagnetic energy into mechanical energy. Motors work from the principle of a rotating dipole. An example is the DC motor. The DC motor consists of the stator, which is the stationary enclosure, and the rotor, which is the rotating center that drives the axle. In its simplest form, the stator is a permanent magnet, which sets up a strong ambient magnetic field. The rotor is basically a coil of wire that forms a magnetic dipole when a DC current is driven through the wire. The rotor acts like a compass needle and moves to align its dipole moment with the magnetic field. To get the rotor to rotate continuously, some ingenious engineering is used. Just before the rotor completely aligns itself with the field, the DC current in the rotor is disconnected. The rotor's angular

EFFECTS OF THE MAGNETIC FIELD

momentum causes it to freely rotate past alignment. Then the DC current is reconnected, but with reverse polarity. The rotor's dipole is consequently reversed, and the rotor is now forced to continue rotating another 180 degrees to try to align with the field. The process repeats ad nauseam. This simple example is called a two-pole motor because the rotor has two poles, north and south. More than one dipole can be used in a radial pattern on the rotor to produce a more powerful motor. You have now learned another way that electrical energy can be taken from a circuit. In Chapter 2, you learned that a resistor is just a device that converts electrical energy into heat. A motor converts electrical energy into mechanical energy. From the point of view of the circuit, this energy loss also appears as a resistance, although there is no "resistor" involved.

The Moving Charge

Another, more fundamental, test particle for the magnetic field is a free charge moving with velocity, v. As you learned earlier in this chapter, the magnetic field arises from moving charges. Therefore, a moving charge serves as a good test particle.

You can better understand the effect that a magnetic field has on a moving charge by first understanding a similar mechanical effect, that of the Coriolis force. Without knowing it, you are probably very familiar with the Coriolis force. Imagine that you are standing on a spinning platform, such as a merry-go-round or a giant turntable. You are standing at the center of the platform, and your friend Bob is standing on the other side. Furthermore, the platform is spinning counterclockwise (as seen from above). You are playing catch and you throw a baseball directly at Bob. To your dismay, the ball does not travel in a straight line to Bob, but curves off to the right. From the perspective of you and Bob, it is as if a force acted on the ball, making it curve to the right. This apparent force is the Coriolis force. (There is also a centrifugal force present, which is discussed briefly later in this chapter.) Although no real force acts on the ball, from the reference frame of the spinning platform, it appears as if a force acts. Figures 3.9 and 3.10 illustrate the situation.

Next, Bob takes a baseball and throws it toward you. This time the ball curves to your left. You have discovered something else. The force depends on the direction of the throw. In fact, it also depends on the speed of the throw. Now what if the platform was spinning at a different rate, or what if the platform changed direction of spin? You can easily convince yourself that both of these changes would affect the

Figure 3.9 You throw a ball to your friend Bob. The platform is not spinning. The ball, therefore, travels straight to Bob.

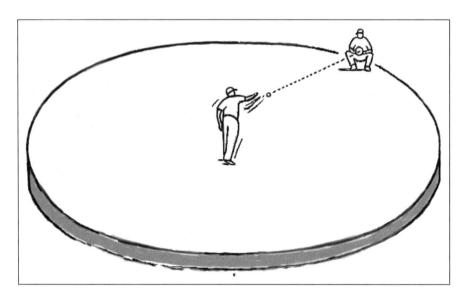

Figure 3.10 You throw a ball to your friend Bob. The platform is spinning, and the ball curves to the right of Bob.

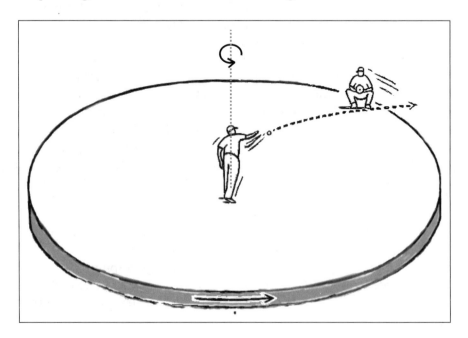

EFFECTS OF THE MAGNETIC FIELD

path of the ball and the apparent force. The exact mathematical formula for the magnitude of the Coriolis force is

$$F_{Coriolis} = 4\pi mvf \sin(\theta)$$

where m is the mass of the ball, v is the speed of the ball, f is the rotation frequency of the platform, and θ is the angle between the rotation axis and the direction of the ball's velocity. The rotation axis is the axis on which the platform is rotating. The direction of the force is determined by the cross product right hand rule, and the full formula for the force is

$$\vec{F}_{Coriolis} = 4\pi m \left(\vec{v} \times \vec{f} \right)$$

where the direction of f is defined as upward for a counterclockwise rotating platform, and downward for a clockwise rotating platform.

Could you throw the ball straight to Bob? No. You could account for the Coriolis force and aim your throw to the left of Bob. But even though the ball might make it to Bob, while it was in the air its flight path would still be curved. You could force the ball to travel straight to Bob if you had a pipe or tube connecting the path between the two of you. The ball would be forced to travel in a straight line because the pipe would provide an equal but opposite force to counteract the Coriolis force.

The Coriolis force causes many effects on earth. The spin of the earth about its axis causes a global Coriolis force, which is responsible for many weather effects on our planet. For instance, the Coriolis force is what causes hurricanes to rotate counterclockwise in the Northern Hemisphere and clockwise in the Southern Hemisphere.

The magnetic force acts in the exact same manner as the Coriolis force. Imagine you now are trying to have the same game of catch with your friend Bob. However, instead of throwing a baseball, you are now throwing a positively charged metal sphere to him, as shown in Figure 3.11. You also happen to be standing in a constant magnetic field whose direction is upward. You throw the ball directly at Bob, but to your amazement it curves off to the right. In fact, it behaves the way the baseball did when you were spinning on the merry-go-round. However, in this case neither of you is spinning! Very strange indeed. Continuing the analogy, to a charged particle, the magnetic field makes space seem like it is rotating, with the rotation taking place about an axis in the direction of the magnetic field.

Figure 3.11 Immersed in a magnetic field, you throw a charged metal ball to your friend Bob. The ball curves to the right of Bob, makes a loop, and eventually returns to you.

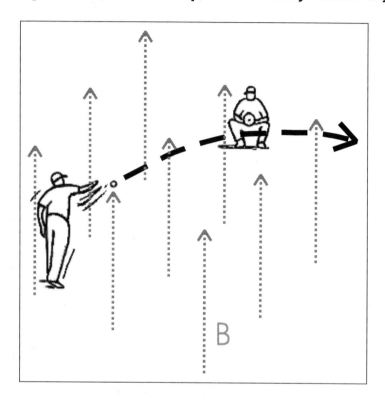

The exact formula for the force is called the Lorentz force law, which is expressed as

$$\vec{F}_{magnetic} = q(\vec{v} \times \vec{B})$$

or, in terms of magnitude only,

$$F_{magnetic} = qvB\sin(\theta)$$

where q is the magnitude of the charge on the moving object, v is the velocity of the object, B is the magnitude of the magnetic field, and θ is the angle between the magnetic field and the direction of the ball's velocity. The direction of the force is again determined by the cross product right hand rule. The Lorentz force law forms the basis for how the magnetic field transmits its action. All magnetic effects can be ultimately reduced to this law.

EFFECTS OF THE MAGNETIC FIELD

There is an interesting side effect to the magnetic force in this example. Assuming that there is no air friction, the charged metal ball will continue to curve forever. Therefore, it will trace out a circular path. Furthermore, the path of the free charge creates a field whose direction is opposite to that of the applied field. The free charge orients so that its dipole moment opposes that of the field. Therefore, the rotation of objects like compass needles and motor coils, which rotate so as to reinforce the field, must stem from something other than free charges.

Although the mechanical example and the magnetic example are very similar, there is one important difference, as illustrated in Figure 3.12. In the mechanical example, there is the additional centrifugal force, which causes the ball to move outward from the center. (The Coriolis and centrifugal forces are actually virtual forces, as your frame of reference, not the ball, is being accelerated.) So the ball thrown from the spinning merry-go-round will appear to spiral away forever. In contrast to this behavior, the ball thrown in the magnetic field behaves like a boomerang, tracing out the same path forever.

Aurora Borealis: The "Northern Lights"

In Figure 3.10, the charge begins its velocity at a direction exactly perpendicular (90 degrees) to the magnetic field. If the velocity is not exactly perpendicular, the charge will follow a helical path along the magnetic field lines. In other words, it moves in the direction of the field lines, as well as encircling them. This result follows directly from the Lorentz magnetic force law. In addition, this phenomenon is responsible for the aurora borealis or "northern lights," the fantastic natural light show seen in the arctic regions. (In Antarctica it is called the "southern lights.") Charged particles (protons and electrons) that are part of the solar wind are swept into the vicinity of the Earth's magnetic field, which extends out past the atmosphere. These particles are caught by the earth's magnetic field, and they tend to spiral around the magnetic field lines toward the north and south poles. As the particles get closer to the polar regions, they descend through the atmosphere. In the atmosphere, they collide with gas atoms (mainly oxygen and nitrogen), causing the atoms to ionize. During the ionization process, some of the particles' kinetic energy is converted to light. This visible light is the aurora borealis.

Currents

Another test particle that can be used in analyzing magnetic fields is the current segment. Keep in mind that there is a distinct difference

Figure 3.12 A) Path of ball thrown in Figure 3.10 (Coriolis force + centrifugal force). B) Path of ball thrown in Figure 3.11 (magnetic force).

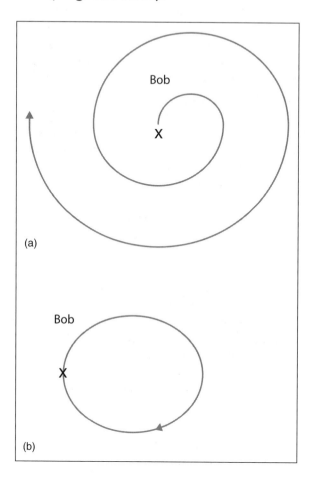

between a moving charge and a current. A current consists of a group of moving charge that occupies a length in space, as opposed to a charge, which occupies only a point in space. The different points of a current are also typically rigidly connected, as is the case with a current in a wire. This point is key. In the case of a free electron, the magnetic field acts at one specific point in space. With a current, the magnetic field acts in many places at the same time, acting to move the entire structure. Second, a current always implies the existence of another force with the job of always keeping the current at its same value. In the typical case of a current in an ordinary wire, this second force is the electric field, which is imposed by the source. Another common difference

Figure 3.13 A current-carrying wire is forced to the right by an external field.

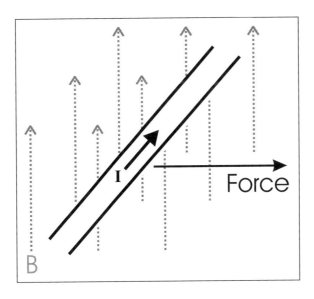

is that most currents occur in wires made of atoms. As opposed to static electricity, the dynamic electricity of electronic circuits involves conductors that have overall neutrality. For every electron flowing in the current, there is a corresponding stationary hole or positive ion (refer back to Chapter 2).

Consider the situation in Figure 3.13. A wire carrying a current is placed in a constant magnetic field that points upward. As in the case of the single charged particle, the charges of the current are initially pulled to the right, dragging the wire with it. However, with the current, we have a second force—the electric field in the wire. Consequently, the charges also continue to move down the wire. The outcome is that the electric field fights against the tendency for the charges to try to circle back as in the case of the single charge. As long as the source for the current continues to drive a constant current, the wire will continue to move to the right. The energy expended to move the wire comes from the current source. The current source must supply an extra amount of electric field to counteract the magnetic force. Again, this energy loss corresponds to the appearance of a resistance from the circuit point of view.

A similar situation occurs when two parallel, current-carrying wires are placed near each other. From Figure 3.14 you can determine that if the wires are carrying current in the same direction, the wires will be attracted to each other, moving together until their magnetic fields

Figure 3.14 Two wires that carry current in the same direction are attracted to each other.

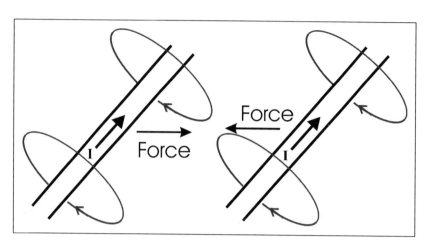

coincide. If the wires carry current in opposite directions, they will be repelled from each other.

For another example, consider a loop of current in a magnetic field as shown in Figure 3.15. By using the cross product right hand rule, you can see that the loop will rotate to align its magnetic dipole with that of the imposed field. You may also note the strange fact that if the dipole is placed exactly opposite to the field, it will not move. The situation is similar to the theoretical fact that a pendulum perfectly balanced at its peak will remain stationary in an inverted position, like a pencil on its point. In reality, any slight deviation from perfect balance will cause the pendulum to fall and eventually settle at its bottom-most position. The same can be said of the magnetic dipole.

At this point, I have come full circle. I introduced the magnetic field by describing how dipoles rotate to line up with the field without explaining why. I then introduced the Lorentz law for magnetic forces and the cross product right hand rule to explain the fundamental effects of magnetism. I then proceeded to explain the reason why the dipole orients itself by using the Lorentz force law.

The general law governing magnetics is that currents and magnets will move so that their magnetic fields line up to produce the minimal energy of the total field. This statement explains why two magnets placed near each other move and rotate until the south pole of one magnet is touching the north pole of the other.

Figure 3.15 A current-carrying loop experiences a torque (rotational force) causing it to line up with an applied magnetic field.

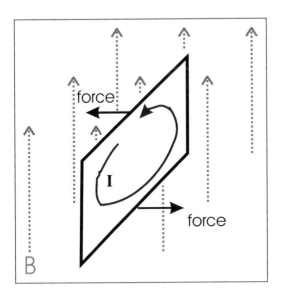

Audio Speakers

Audio speakers are another practical application of magnetics. In the typical speaker, an electromagnet consisting of a coil is placed in line with a permanent magnetic. The cone of the speaker is then attached to the coil, while the permanent magnet is held fixed in place. Depending on the direction of the current, the coil will either be pulled toward the permanent magnetic or repelled away from the permanent magnetic. The AC signals of music or speech cause the speaker coil and cone to vibrate in concert with the signal. The vibrations create sound waves in the surrounding air and you hear the signal. The energy expended to create the sound appears as resistance from the circuit point of view.

Here's a question for you: Why do speakers have plus and minus terminals on them? The speaker is not grounded. The signals are AC, so only the relative motion matters. The answer has nothing to with electronics. Instead this requirement is because of acoustics. The two speakers of a stereo system must have the same polarity so the sound created by the speakers adds constructively. If you mismatch the polarity of one speaker, you will get a dead zone of sound between the two

68 FUNDAMENTALS OF MAGNETIC FIELDS

speakers. If you match the polarity of both speakers, your stereo will sound fine.

THE VECTOR MAGNETIC POTENTIAL AND POTENTIAL MOMENTUM

In the previous chapter, which covered electric fields, one of the first concepts covered was the electric field potential, more commonly known as voltage. You may be wondering if a similar potential exists for the magnetic field. If so, you are correct. However, the magnetic potential is a vector quantity. It has both magnitude and direction. The vector potential around a current is shown in Figure 3.16. As you can see, its main characteristic is that it points in a direction parallel to the current, and it decays in magnitude as the distance to the current increases.

The magnetic vector potential is much harder to understand than voltage, the electric potential. However, I will sketch out some of its characteristics. The magnetic field stores energy just as the electric field stores energy. In some situations the vector potential can be interpreted

Figure 3.16 A plot of the magnetic vector potential surrounding a current-carrying wire.

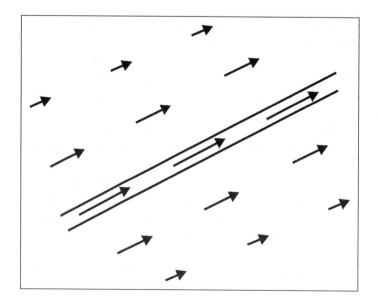

as the potential momentum of a charge. In fact, the units of the vector potential are those of momentum per charge. When Maxwell developed his theory of electromagnetism, he called the vector magnetic potential the "electrodynamic momentum" because it can be used to calculate the total momentum or total kinetic energy of a system of charged particles and their electromagnetic fields. In Chapter 6, you will learn more about the vector potential when we discuss quantum physics.

MAGNETIC MATERIALS

Diamagnetism

In Chapter 2, you learned that different materials behave differently in electric fields. You learned about conductors and dielectrics. Electric fields induce reactions in materials. In conductors, charges separate and nullify the field within the conductor. In dielectrics, atoms or molecules rotate or polarize to reduce the field. Magnetic fields also induce reactions in materials. However, since there are no magnet charges, there is no such thing as a "magnetic conductor." All materials react to magnetic fields similarly to the way dielectrics react to electric fields. To be precise, magnetic materials usually interact with an external magnetic field via dipole rotations at the atomic level. For a simple explanation, you can think of an atom as a dense positive nucleus with light electrons orbiting the nucleus, an arrangement reminiscent of the planets orbiting the sun in the solar system. Another similar situation is that of a person swinging a ball at the end of a string. In each situation, the object is held in orbit by a force that points toward the orbit center. This type of force is called a centripetal force. The force is conveyed by electricity, gravity, or the string tension, respectively, for the three situations. Referring to Figure 3.17 and using the cross product right hand rule, you find that the force due to the external magnetic field points inward, adding to the centripetal force. The increase in speed increases the electron's magnetic field, which is opposite to the external field. The net effect is that the orbiting electron tends to cancel part of the external field. Just as the free electron rotates in opposition to a magnetic field, the orbiting electron changes to oppose the magnetic field. This effect is called *diamagnetism* and is just like that of dielectrics, where the dielectrics tend to reduce the applied electric field. The major difference here is that diamagnetism is an extremely weak effect. Even though all materials exhibit diamagnetism, the effect is so weak that you can

Figure 3.17 A) A circulating electron whose magnetic moment points upward. B) Applying a downward magnetic field increases the upward moment of the electron. (Recall that an electron has negative charge.)

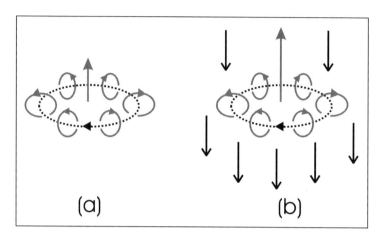

usually ignore it. This explains the commonly known fact that most materials are not affected by magnets.

Paramagnetism

In addition to orbiting the atom, each electron has a spin, which can be thought of in simple terms as similar to the Earth spinning on its axis. Because the electron has a net charge, the spin causes a circular current and a magnetic dipole. We learned earlier that although currents are governed by the same magnetic law as free electrons, they behave in opposite ways. Therefore, the inherent dipole of the electron will be rotated to line up with the external magnetic field, thereby increasing the overall field. Paramagnetism,* while stronger than diamagnetism, is another very weak effect and can usually be ignored. The reason for its weakness is that the electrons in each atom are always grouped in pairs that spin opposite to one another. Hence, paramagnetism can only occur in atoms that have an odd number of electrons. For example, aluminum has an atomic number of 13 and thus has an odd number of electrons. It therefore exhibits paramagnetic properties. The random

---

*There is a simple mnemonic that can be used to recall the definitions of paramagnetism and diamagnetism. With *para*magnetism, the dipoles line up *para*llel to the external field. With *dia*magnetism, the dipoles line up *dia*metrically opposed to the external field.

thermal motions of the atoms tend to prevent the dipole moments from lining up well, even when exposed to an external field.

Ferromagnetism and Magnets

Diamagnetism and paramagnetism are both rather obscure phenomena to most engineers. The effects of ferromagnetism, however, are quite pronounced and quite well known. Ferromagnetism is responsible for the existence of magnets—that is, permanent magnets like the ones you use to hang up notes on the refrigerator. A ferromagnetic material is like a paramagnetic material with the added feature of "domains." Each domain is a microscopic patch of billions of atoms that have all lined up their dipole moments in the same direction. It so happens that the quantum mechanical properties of certain materials, notably iron, cause these domains to form spontaneously. This is due to the electron spin and to the collective behavior of the outermost electrons of large groups of atoms. Normally, the domains are randomly oriented so that the material still has no overall magnetic dipole. However, when a magnetic field is applied the domains that align to the field grow, while domains of other orientations shrink. In addition, the domains have a tendency to freeze in place after aligning. In other words, ferromagnetic materials have memory. For example, if a bar of iron is placed in a strong magnetic field and then removed, the bar retains a net magnetic dipole moment. It has become a magnet. Some other examples of ferromagnetic materials are nickel, which is used in guitar strings, and cobalt. Several metal alloys are also ferromagnetic.

Incidentally, there is also a ferroelectric effect. Ferroelectric materials tend to retain an electric field after being exposed to a large electric field. Analogous to the term "magnet," the term "electret" is used for objects that exhibit ferroelectricity. Electrets are used in certain types of microphones.

Demagnetizing: Erasing a Magnet

There are two common ways to demagnetize (to remove any net magnetic field in) a magnet. First, you can heat the magnetic past its "Curie point." Just as ice melts to water, the frozen magnetic moment will "melt away" because above this temperature the material is no longer ferromagnetic. The second and more practical technique is to expose it to the strong AC magnetic field of an electromagnet, such as a solenoid. The field is then slowly decreased to zero. By the end of the process, the object will have a negligible magnetic field. This technique is known as "degaussing."

Table 3.1 Magnetic Classification of Materials

Material Type	Description
Nonmagnetic	No magnetic reaction.
Diamagnetic	Induced dipole moment *opposes* applied field. Repelled by bar magnet. Very weakly magnetic.
Paramagnetic	Induced dipole moment aligns to applied field. Attracted by bar magnet. Weakly magnetic.
Ferromagnetic	Induced dipole moment aligns to applied field. Attracted by bar magnet. Very strongly magnetic. Has *memory* and so can be used to create permanent magnets. High electrical conductivity.
Ferrimagnetic	Type of ferromagnetic material. Induced dipole moment aligns to applied field. Attracted by bar magnet. Very strongly magnetic.
Ferrites	Type of ferrimagnetic material. Induced dipole moment aligns to applied field. Attracted by bar magnet. Very strongly magnetic. *Low electrical conductivity.*
Superparamagnetic	Material mixture: ferromagnetic particles suspended in a plastic binder. Induced dipole moment aligns to applied field. Very strongly magnetic. Has memory, which allows for uses in audio, video, and data recording.

Data adapted from Krauss and Fleisch, *Electromagnetics with Applications*, 5th Edition, McGraw-Hill, 1999.

Summary of Magnetic Materials

In summary, some magnetic materials line up with an external field and some materials line up opposite to the field. This result is similar to the way free electrons line up to oppose a field, whereas controlled currents move and/or rotate to reinforce a field. Table 3.1 summarizes some of the types of magnetic materials.

Table 3.2 Low-Frequency Permeability of Various Materials

Material	Type	Permeability
Copper	Diamagnetic	$0.999991 \times \mu_0$
Water	Diamagnetic	$0.999991 \times \mu_0$
Vacuum	Nonmagnetic	$1 \times \mu_0$
Air	Paramagnetic	$1.0000004 \times \mu_0$
Aluminum	Paramagnetic	$1.00002 \times \mu_0$
Nickel	Ferromagnetic	600
Ferroxcube 3, Mn-An Ferrite Powder	Ferromagnetic (Ferrite)	1500
Iron (0.2% impurities)	Ferromagnetic	5000
Iron (0.05% impurities)	Ferromagnetic	200,000 *

$\mu_0 = 4\pi \times 10^{-7}\,\text{H/m}$

Data adapted from Krauss and Fleisch, *Electromagnetics with Applications*, 5th Edition, McGraw-Hill, 1999.

Table 3.2 gives the properties of a few magnetic materials. The relative permeability of each material is given. Permeability quantifies how a material responds to magnetic fields in a manner analogous to how permittivity quantifies the material response to an electric field. It follows that permeability is a measure of the magnetic energy storage capabilities of material. A material with a relative permeability of 1 is magnetically identical to a vacuum, and therefore stores no magnetic energy. Paramagnetic and ferromagnetic materials have relative permeability greater than 1, which implies that the material aligns its dipole moments to an induced field and therefore stores energy. Higher permeability translates to a larger reaction and higher energy storage. Diamagnetic materials are characterized by relative permeabilities less than 1. This fact implies that the material aligns its dipole moments opposite to an induced field. Because the material reacts to the field, it also stores energy. Hence a lower permeability for a diamagnetic material translates to higher energy storage.

MAGNETISM AND QUANTUM PHYSICS

This chapter's descriptions of diamagnetism, paramagnetism, and ferromagnetism are only approximations. To truly explain these effects, quantum mechanics and solid state physics (quantum theory of solids) are necessary. In quantum physics, the electron, due to its inherent wave nature, acts more like a spread-out glob engulfing the nucleus rather than a miniature planet. Furthermore, in quantum mechanics the electron's spin is more of a theoretical quantity and can occur in only one of two quantum states, up or down. You will learn more about quantum physics in Chapter 6.

BIBLIOGRAPHY

Blatt, F. J., *Principles of Physics*, 3rd Edition, Boston, Mass.: Allyn and Bacon, 1989.

Cogdell, J. R., *Foundations of Electrical Engineering*, 2nd Edition, Englewood Cliffs, N.J.: Prentice-Hall, 1995.

Encyclopedia Britannica Inc., "Magnetism"; "Aurora," *Encyclopedia Britannica*, Chicago: Encyclopedia Britannica Inc., 1999.

Epstein, L. C., *Thinking Physics—Is Gedanken Physics; Practical Lessons in Critical Thinking*, 2nd Edition, San Francisco, Calif.: Insight Press, 1989.

Feynman, R. P., R. B. Leighton, and M. Sands, *The Feynman Lectures on Physics Vol II: Mainly Electromagnetism and Matter*, Reading, Mass.: Addison-Wesley Publishing, 1964.

Fowles, G. R., and G. L. Cassiday, *Analytical Mechanics*, 6th Edition, Fort Worth, Texas: Saunders College Publishing, 1999.

Griffiths, D. J., *Introduction to Electrodynamics*, 3rd Edition, Upper Saddle River, NJ: Prentice Hall, 1999.

Halliday, D., R. Resnick, and J. Walker, *Fundamentals of Physics*, 6th Edition, New York: John Wiley & Sons, 2000.

Kittel, C., *Introduction to Solid State Physics*, 7th Edition, New York: John Wiley, 1996.

Kraus, J. D., and D. A. Fleisch, *Electromagnetics with Applications*, 5th Edition, Boston: McGraw-Hill, 1999.

Purcell, E. M., *Electricity and Magnetism*, Boston, Mass.: McGraw-Hill, 1985.

4 ELECTRODYNAMICS

Now that you have learned separately about the electric and magnetic fields, it is time to learn about the effects that can be explained only by taking into account both the electric and magnetic field. In other words, it is time to learn about the interaction of the two fields.

CHANGING MAGNETIC FIELDS AND LENZ'S LAW

In Chapter 3 you learned about the magnetic field and how it stores energy. You may wonder where this magnetic field energy comes from. The energy of the magnetic field comes from the current that caused it. Consider a simple electric circuit where a power source supplies a resistor through some wire. When the power supply is turned on, the current in the circuit increases from zero to that defined by Ohm's law, $V = I \times R$. In the steady state, a constant magnetic field surrounds the wire, as described in Chapter 3. This field stores energy, which must somehow have been transferred from the power supply. The energy transfer is governed by Lenz's law. Lenz's law is the magnetic corollary to Newton's third law: for every action there is an equal and opposite reaction. Loosely put, Lenz's law states that whenever a change in a magnetic field occurs, an electric field is generated to oppose the change. The opposing electric field is sometimes called *back-emf*, where emf stands for electromotive force.

Continuing with the present example, when the power is turned on, a magnetic field starts to develop around the wires. Because the magnetic field is changing from zero to a non-zero value, Lenz's law states that an electric field is generated that opposes the change. This electric field manifests itself in the circuit as voltage. The opposing voltage persists until the current reaches its final steady-state value. The current, therefore, cannot change instantaneously, but continuously changes from zero to its final value over a period of time. Furthermore, while the current is increasing, a voltage drop exists across the wires. A voltage together with a current implies power loss. Although all real wires have resistive (heating) losses, you can ignore such losses for this example. The power loss encountered here actually corresponds to the power transferred into the magnetic field surrounding wires. Just as it takes energy to increase the speed of a car, it also takes energy to increase the speed of change in a circuit (i.e., the current). You can think of Lenz's law as a way in which nature "balances its books." Energy is always conserved, and Lenz's law tells us how energy conservation is maintained with magnetic fields.

FARADAY'S LAW

Lenz's law provides a qualitative understanding of how a changing magnetic field creates an electric field. Faraday's law, proposed by 19th-century scientist Michael Faraday, describes this action qualitatively. For a solenoid with N turns and a cross sectional area A, Faraday's law can be written as

$$V = -A \cdot N \frac{dB}{dt}$$

where dB/dt is the change in magnetic field per unit time and V is the resulting voltage in the circuit.

INDUCTORS

At last it is time to learn about inductors. In contrast to the capacitor, which requires only one field (namely the electric field) to describe its operation, the inductor requires both fields to describe its operation even though it stores only magnetic energy. Here again is an inherent difference between the electric and magnetic fields. An inductor is a circuit element used to store magnetic energy. Typically, an inductor is created from several loops of wire stacked together to form a solenoid. The stacking of several loops serves to concentrate a large magnetic field in a small volume. The magnitude of an inductor is measured by its inductance, which depends on the size and shape of the coil of wires. For a very long solenoid, the inductance is approximately

$$L = \mu_0 \cdot N^2 \frac{A}{\ell}$$

where μ_0 is the permeability of free space, N is the number of turns of wire in the solenoid, A is the cross-sectional area of the solenoid, and ℓ is the length of the solenoid. Larger inductance translates to larger magnetic energy storage for a given current. More explicitly, the magnetic field inside an inductor is related to the driving current as

$$B = \frac{LI}{NA}$$

Discharging an Inductor

Upon learning that the circuit supplies energy to the magnetic field of an inductor, you may wonder where this energy goes when the circuit is turned off. The short answer is that the energy is returned to the circuit. For example, imagine a simple R–L (resistor and inductor in series) circuit powered by a DC source. After the initial charging period, a steady current exists in the circuit, and a steady magnetic field exists inside the inductor. Now imagine that someone opens the circuit. The current in the circuit is abruptly brought to zero. Invoking Lenz's law, you know that a voltage is created to counteract the change in current. The inductor "wants" to continue pumping current, just as a car "wants" to continuing moving once set in motion. In other words, both the car and inductor have some form of inertia. Even though the circuit is broken, the magnetic field will continue to pump current through the inductor. Since the charge has no circuit to follow, it builds up at the ends of the inductor. A large voltage results. The "charged" inductor acts like a current source, analogously to the charged capacitor, which acts as a voltage source. In an ideal inductor, the current would continue to flow indefinitely, leading to an infinite voltage. In the same way, an ideal car (i.e., a car with no frictional forces), once set in motion, would roll forever, as stated by Newton's first law. In any real circuit, however, infinite voltages are impossible. When the voltage reaches a high enough value, the air between the contacts of the switch will break down and a spark occurs. The spark forms a temporary path for current, allowing the inductor to discharge its energy to the resistor of the circuit. This is essentially how a spark is generated in the ignition system of a car. The spark occurs in a small gap between two electrodes of the spark plug, igniting the vaporized gas inside the piston. The turning of the engine opens and closes the switch that charges the coil.

Although a spark is the most dramatic result of discharging an inductor, the energy can be dissipated less violently through the parasitic elements of the inductor. All real inductors have a parasitic capacitance that forms between the coils of the solenoid. The parasitic capacitance couples with the inductance to form a resonant circuit. The voltage in the inductor reaches a peak value when the capacitance is fully charged. Then the capacitance discharges current into the inductor. The process repeats itself and a sinusoidal oscillation is produced. As the energy cycles between the inductance and capacitance, it gradually diminishes because the parasitic resistance of the coil wire converts it to heat. You will learn more about parasitic elements in Chapter 7.

So far I have described the ways an inductor loses its magnetic energy. Can an inductor be used as a long-term storehouse of energy like a capacitor? Yes, but it ain't easy. In the case of a capacitor, you just have to remove the capacitor from its circuit and it will continue to hold its voltage. In other words, you must place an open circuit across a capacitor for it to store its energy. It follows that since an inductor is the electromagnetic dual to a capacitor, you must place a short circuit across an inductor for it to store its energy. In practice this can only be accomplished through the use of superconductors. Any ordinary wire will quickly sap the inductor's energy because of its resistance. Nature has made it much easier for us to construct a low-loss open circuit than to construct a low-loss short circuit.

AC CIRCUITS, IMPEDANCE, AND REACTANCE

An interesting thing happens when you apply an AC source to an inductor. Power is transferred to and from the inductor during each cycle, and the energy provided by the circuit is equal to the energy returned from the inductor. Because energy is not lost, the ideal inductor does not have an AC resistance. However, sinusoidal current flows through the inductor and a sinusoidal voltage develops across the inductor. In other words, the ratio of voltage to current is not zero. Moreover, the current wave and the voltage wave are always 90 degrees out of phase. Through the use of complex numbers, impedance (the generalization of resistance) can be defined. Impedance, Z, is the ratio of voltage to current in any circuit element. Real impedances imply resistance and therefore power lost from the circuit. Imaginary impedances imply energy storage in a circuit. Imaginary impedance is called *reactance* because it is a reaction to the voltage source. Ohm's law can be extended with the concept of impedance and complex numbers,

$$V = I \cdot Z$$

where V and I are the root-mean-square voltage and current (in phasor notation) respectively.

In an inductor, the current wave lags the voltage wave by 90 degrees and its impedance is defined as

$$Z = i2\pi f L$$

where f is the wave frequency and i is the square root of -1. Capacitors also store energy, and therefore cycle energy in AC circuits. In a capacitor, the current wave leads the voltage wave by 90 degrees and its impedance is defined by

$$Z = \frac{-i}{2\pi f C}$$

From these definitions, you can see that positive imaginary impedance implies inductive reactance and negative imaginary impedance implies capacitive reactance.

RELAYS, DOORBELLS, AND PHONE RINGERS

When an object is placed in a magnet field, its atomic dipole moments react as described in Chapter 3. Usually this effect is negligible, except for certain ferromagnetic materials such as iron, nickel, and steel. When a piece of iron is placed in a magnetic field, its internal atomic dipoles align with the field. The object will be attracted to the source of the field, and vice versa. We encounter this phenomenon whenever we place a magnet on the refrigerator door. This is the magnetic corollary to static electricity, which causes dust to stick to TV screens and static-charged clothes to stick together.

This "static magnetism" is not limited to permanent magnets—ferromagnetic objects are also attracted to electromagnetics. An iron bar will be attracted to and drawn into a solenoid. This mechanism is exploited in many electromechanical devices. For instance, relays are switches controlled by current in a solenoid. An iron plunger normally holds the switch in one position with the aid of a spring. When enough current is sent through the solenoid, the magnet field pulls the iron plunger into the solenoid with enough force to counteract that of the spring. The switch position is then changed. The same mechanism is used for doorbell ringers and for the ringers of old telephones. To ring your telephone, the phone company sends a large (~100V) signal on your telephone line at a frequency of about 20 Hz. In an old-fashioned phone, the large AC signal causes a large AC magnetic field in a solenoid. An iron bar inside the solenoid is made to vibrate back and forth. The iron bar is attached to a clapper, which hits a bell at each cycle of the wave. A doorbell ringer works on the same principle.

MOVING MAGNETS AND ELECTRIC GUITARS

Now instead of letting the iron bar and solenoid move together, imagine that you hold the solenoid fixed and move the piece of iron back and forth. By moving the iron bar, you are also moving the induced magnetic field of the iron bar. From the frame of reference of the stationary solenoid, the moving iron bar creates a moving magnetic field. From Lenz's law you know that an electric field will arise to counteract the changing magnetic field. The induced electric field creates a voltage in the solenoid. This is one of the exact experiments Faraday performed, leading to the law that bears his name. The important consequence is that mechanical energy (moving a bar of iron) can be converted to electrical energy (a voltage in the solenoid). This result is the basis for electrical generators and for electric guitars.

The electric guitar converts the vibrating energy of the nickel strings into an AC current. Under the strings, near where the guitarist strums, a solenoidal device called a pickup is placed. The pickup typically consists of six permanent magnet cylinders, one placed under each guitar string. A single coil of wire is wrapped around all the permanent magnets, forming a solenoid with six iron cores. The magnetic field of the permanent magnets induces a secondary magnetic field in the guitar strings because the strings are made of a ferromagnetic material. When a string is plucked, it vibrates at its resonant frequency. Its magnetic field moves along with it. The moving field induces a voltage in the solenoid below, as required by Lenz's law. The overall effect is that of six AC voltage sources (one for each string) connected in parallel.

GENERATORS AND MICROPHONES

Generators and microphones work on the same principle of Faraday's law. Any electronics hobbyist knows that a DC motor can also be used as a DC generator, and a speaker can also be used as a microphone. Consider the DC motor described in Chapter 3. In Chapter 3, you learned that applying a voltage across the rotor coil caused it to rotate because of the permanent magnetic field of the stator. If, instead, you move the rotor by a mechanical means, the orientation of the rotor will change with respect to the magnetic field of the stator. A voltage is thereby induced in the rotor. Mechanical energy is converted to electrical energy. Faraday was in fact the first person to create an electric generator. When he demonstrated his device to the prime minister of England, the prime minister asked what this device could be used for. Faraday is said to have replied sarcastically, "I know not, but I wager that one day your government will tax it," or least that is how the story goes. Of course Faraday was correct, as we all know too well. Faraday also knew that generators and motors would be of great utility for harnessing and transferring energy.

Returning again to Chapter 3, you learned how an electromagnet can be used to cause a paper cone to vibrate, producing sound. If, instead, the cone is forced to vibrate by incoming sound, an electric field is induced in the coil by means of the moving permanent magnet attached to the coil. In this way, sound waves such as those produced by the human voice can be converted to electric signals.

THE TRANSFORMER

The transformer also relies on Faraday's law for its operation. A transformer is constructed from two solenoid inductors such that the magnetic fields of each inductor are closely coupled. The inductors can be coupled closely by stacking them on top of one another. By varying the current in one of the inductors an AC magnetic field is created. By means of Faraday's law, a voltage is induced in the other inductor. In this way the transformer can be used to transfer energy between two unconnected circuits.

Transformer and Inductor Cores

Another, more efficient, method for coupling inductors is the use of a core material with a high permeability, typically a ferromagnetic material such as iron. The core acts to concentrate the magnetic field so most of the field is coupled between the inductors. Ferromagnetic cores can also be used to increase the inductance of inductors. To achieve the highest inductance and the best coupling in a transformer, you should use a core that forms a closed loop, such as a toroidal or donut-shaped core. By using such a closed loop core, all of the magnetic field will be concentrated in the core material, not in the air. A side benefit of using a closed loop core is that such components are unlikely to cause interference with other components. With almost all the magnetic field contained in the core, there is not much of a stray magnetic field to induce unwanted effects in nearby components or circuits.

SATURATION AND HYSTERESIS

Unfortunately, with all the usefulness of ferromagnetic cores comes a few unwanted side effects. The first side effect is saturation. Saturation occurs when the applied magnetic field is so large that the permeability starts to drop. If the field is increased further past saturation, the permeability eventually drops to that of free space. When in saturation, the material can no longer increase its internal microscopic magnetic alignment; in other words, the microscopic magnets are fully aligned. An inductor whose core is in saturation will have its inductance drop dramatically. An AC signal that encounters saturation will be clipped or distorted in some way. It is therefore important in most circuit applications to prevent the magnetic field inside the inductor from reaching the level of saturation.

Hysteresis is another effect that occurs in ferromagnetic cores. Hysteresis is associated with the inherent memory of ferromagnetic materials that allows for the creation of permanent magnets. After a magnetic field is applied to a ferromagnet, some of the magnetization is retained, even after the applied field is turned off. So even though you brought the applied field back to its original starting point, the material has not returned to its original state. At the microscopic level, there is a force that tends to oppose the realignment of the microscopic magnets. The dependence of the material's state on its past history is known as *hysteresis*. In addition to memory effects, hysteresis also causes energy loss. When you apply a magnetic field that realigns the microscopic magnets, a frictional effect occurs, and energy is lost to heat.

WHEN TO GAP YOUR CORES

Earlier in the chapter, I mentioned that to obtain the highest inductance and smallest amount of stray magnetic fields, a ferromagnetic core in the form of a loop is used. To reach a compromise between the benefits and disadvantages of ferromagnetic cores, an air gap is often placed within the core loop. In such a configuration, the overall permeability takes on an intermediate value between that of the core material and that of air. The effects of hysteresis are also mitigated because air has no hysteresis.

In addition, with the gap of air in series with the ferromagnetic core, most of the magnetic energy gets stored in the air. This may seem counterintuitive, but it is analogous to how resistors behave in electric circuits. If you place a large resistor in series with a small resistor, the larger voltage drop (and larger energy loss) will occur across the large resistor. Conversely, when the resistors are connected in parallel, the larger current (and larger energy loss) will occur in the small resistor. Continuing the analogy, when air is placed in parallel with iron, most of the magnetic field (and energy storage) occurs in the iron.

So when should you gap and when should you not gap the core of a transformer? This question usually arises in the context of power supply design, where the goal of the engineer is to output a large amount of power at the lowest material cost, and often, the lowest overall weight and size. To get the most "bang for your buck," you should always use an ungapped transformer whenever possible because it provides the highest inductance for any given form factor. Thus when a transformer is being used to transfer an AC signal from the source to load, ungapped cores are not usually a problem. However, any situation where the transformer is used to store and then release energy, such as in the use of an ignition coil or a flyback transformer, usually requires a gap in the core to lessen the hysteresis-related losses. In these applications, the primary of the transformer is charged with energy by driving it with a DC current. When the current is turned off, a transient magnetic field occurs. The transient magnetic field causes a voltage to develop across the winding of the secondary, and thus the energy stored in the primary is transferred to the load of the secondary. With a gap placed in the core, most of the magnetic energy is stored in the air gap, greatly reducing hysteresis losses. As an alternative to using a gapped core, you can also use certain porous materials. The porosity of the material creates the effect of a distributed air gap, reducing the hysteresis effects.

FERRITES: THE FRIENDS OF RF, HIGH-SPEED DIGITAL, AND MICROWAVE ENGINEERS

Ferrites are a special subset of ferromagnetic materials. (For a more exact definition, refer to Table 3.1.) Ferrites are a ceramic material, as opposed to other ferromagnetic materials, which are metals. Hence, whereas most ferromagnetic materials such as iron are very conductive, ferrite materials have very poor conductivity. Conductive materials such as iron can be very problematic when used in circuit design. The problems arise from eddy currents. *Eddy currents* are unwanted currents induced in a conductor by a changing magnetic field.

The eddy currents are true currents in that free electrons are actually moving within the material. Associated with currents are resistive losses. Eddy currents can produce high losses inside transformers and inductors, and these losses increase dramatically with frequency. Several techniques can be employed to avoid this problem. One technique is to use bundles of electrically insulated iron rods for a core. The permeability stays about the same, but eddy currents are unable to circulate. This is an old technique useful for low-frequency, high-power applications.

Another option is the use of a ferrite core. Being poor conductors, ferrites do not have the acute problem of eddy currents. Ferrites are the choice core material for high frequencies. However, although ferrites can remedy the eddy current problem, they can't eliminate all loss. At frequencies in the MHz to 100 MHz range, even ferrites become lossy due to heating from other magnetic effects. Therefore, at UHF and microwave frequencies, only air-core inductors are typically used.

Even the high-frequency lossy quality of ferrites has a useful application. Ferrites act as a combination inductor and frequency-dependent resistor whose resistance is proportional to frequency. For this reason ferrite beads and ferrite molds are great for eliminating high-frequency noise on (low-current) power supplies, digital clock signals, and cable signals.

Although not useful for inductors at microwave frequencies, ferrites are used extensively in microwave systems. At microwave frequencies, very-high-resistance ferrites are used for many specialty applications. The ferrite is used with a DC magnetic field applied to it. (The DC magnetic field can be supplied by permanent magnets or by an electromagnet.) With a DC field applied, ferrites become *anisotropic*; that is, their magnetic properties are different in different directions. Simply stated, the DC field causes the ferrite to be saturated in the direction of the field while remaining unsaturated in the other two directions. Voltage-controlled phase-shifters and filters as well as exotic directional devices such as gyrators, isolators, and circulators can be created with ferrites in the microwave region.

MAXWELL'S EQUATIONS AND THE DISPLACEMENT CURRENT

In the 1860s, the British physicist James Clerk Maxwell set himself to the task of completely and concisely writing all the known laws of electricity and magnetism. During this exercise, Maxwell noticed a mathematical inconsistency in Ampere's law. Recall that Ampere's law predicts that a magnetic field surrounds all electric currents. To fix the problem, Maxwell proposed that not only do electric currents, the

MAXWELL'S EQUATIONS AND THE DISPLACEMENT CURRENT

movement of charge, produce magnetic fields, but changing electric fields also produce magnetic fields. In other words, you do not necessarily need a charge to produce a magnetic field. For instance, when a capacitor is charging, there exists a changing electric field between the two plates. When an AC voltage is applied to a capacitor, the constant charging and discharging leads to current going to and from the plates. As I discussed in Chapter 1, although no current ever travels between the plates, the storing of opposite charges on the plates gives the perceived effect of a current traveling through the capacitor. This virtual current is called *displacement current*, named so because the virtual current arises from the displacement of charge at the plates. Maxwell's hypothesis stated that a changing displacement current, such as that which occurs between a capacitor's plates, produces a magnetic field.

And Maxwell Said, "Let There Be Light"

In addition to being a physicist, Maxwell was also an extraordinarily talented mathematician. When he added his new term to the existing equations for electricity and magnetism, he quickly noticed that the mathematics implied that propagating electromagnetic waves could be created. Such waves had never been observed. This was a monumental discovery. Furthermore, when he performed the mathematical derivations, the speed of these new waves was predicted to be that of the speed of light! Before Maxwell, there had been no reason to believe that light and electromagnetism had anything to do with one another. But he had found an unusual coincidence—too much of a coincidence to be ignored. (At the time of Maxwell's prediction, the speed of light had been measured many times and was known to an accuracy of about 5%.)

Maxwell first published his results in 1864, but it wasn't until years later that his theory was accepted by physicists in general. In 1887, eight years after Maxwell's death, Heinrich Hertz confirmed Maxwell's prediction by transmitting an electrical signal through the air. Hertz used a coil of wire in series with a small air gap. Using a high voltage, he forced a spark to be generated across the gap. At a moderate distance he placed an identical coil in series with an identical small air gap. You can imagine Hertz's excitement when he saw that a spark was produced in the second coil whenever a spark occurred in the first coil. Later experiments showed that these "radio waves" behaved just like light waves, except they had a much longer wavelength. At this point, scientists were convinced that light must be a form of electromagnetic oscillation, and

Maxwell was vindicated. The full equations of electromagnetics still bear his name.

Within a decade, in 1896, Guglielmo Marconi demonstrated that messages could be sent over great distances using these newly discovered radio waves. It was in this year that he sent a radio message over a distance of 10 miles. In 1901, he sent a message across the Atlantic Ocean, which brought him great fame. Nikola Tesla also invented and patented a wireless transmission system, and at an earlier date than Marconi's. However, Tesla was involved in many other electrical projects and was not a dynamic public figure like Marconi. Tesla lost the patent lawsuits concerning radio, but, ironically, the court ruling was overturned in the 1943, giving credit to Tesla. Since the work of each man was independent, both Tesla and Marconi should be given credit as the inventors of wireless transmission. However, Marconi was the pioneer of radio engineering, while Tesla focussed on engineering AC power systems.

At the end of the 19th century there was only one major question remaining to be determined concerning electromagnetics. What was the substance that the electromagnetic waves traveled through? All waves were known to be the transmission of disturbances in a medium. Water waves are variations in the surface tension of water. Sound is the variation in pressure of air. Vibrating guitar strings are variations in tension of the string. Scientists of the time assumed that electric and magnetic fields must be a type of tension in some unknown material. They called this fictitious material "aether." We know now that there is no aether. Light is unlike other waves in that it doesn't need a medium for transmission. The aether concept, however, did not die easily. It took Einstein's theory of relativity to resolve the issue.

PERPETUAL MOTION

Conceptually, it is fairly easy to understand how waves are a consequence of Maxwell's addition to Ampere's law. His addition states that a change in an electric field inevitably causes a magnetic field to be created. From Faraday's law you know that a magnetic field developing will cause an electric field to be created. Changing electric fields beget changing magnetic fields, which beget changing electric fields, and so on. When either a magnetic or electric field changes, a sequence of perpetual events is initiated. This changing field propagates away from its source as a wave. The wave propagates forever, or until its energy is absorbed by matter somewhere. The fact that light waves, radio waves, and X-rays are all initiated by this same phenomenon was an amazing fact in Maxwell's time, and it is still quite an amazing fact of nature today.

WHAT ABOUT D AND H? THE CONSTITUTIVE RELATIONS

If you have taken a course on electromagnetics, you will have encountered two other field quantities, namely D and H. D is called the electric displacement field and H is simply called the H-field. (Due to convention, some authors misleadingly refer to H as the magnetic field and B as magnetic flux density or magnetic induction.) So what are D and H? D and H are theoretical quantities used for problems involving macroscopic materials. At the microscopic level there are no D and H. The D field is a combination of the E field added with its electrical reaction in the material. The H field is a combination of the B field added with its magnetic reaction in the material. More explicitly, D and H measure the statistical average in space and time of atomic reactions to the imposed fields. D and H are thus macroscopic quantities. Referring to Chapter 2, $D = \varepsilon E$ and referring to Chapter 3, $B = \mu H$, where ε is the material permittivity and μ is the material permeability. To be precise, H is really a function of B, but convention has us write the relation the other way around. These relationships are referred to as the *constitutive relations*. The third of these relations is $J = \sigma E$, where J is current density or current per area, and σ is the material conductivity. This third equation is just another way of stating Ohm's law. It describes the average movement of electrons in a conductor, as discussed in Chapter 2. The constituitive relationships do not express laws of their own. They just define how materials behave in electric and magnetic fields. Keep in mind that these relationships are not always simple or linear. For example, the memory-like effect of ferromagnetic materials implies that H is a nonlinear function of B (or vice versa if you go by convention).

In the language of thermodynamics and stastical physics, the three constituitive relations are "equations of state." An equation of state is an equation that relates macroscopic variables of a material or system. The most well known equation of state is the Ideal Gas law,

$$PV = NkT$$

where P, V, and T are the pressure, volume, and temperature of the gas, N is the number of gas molecules, and K is Boltzmann's constant. At the microscopic level, the gas consists of molecules randomly moving and colliding with each other. However, we typically measure the macroscopic variables of the gas such as pressure and temperature.

BIBLIOGRAPHY

Blatt, F. J., *Principles of Physics*, 3rd Edition, Boston, Mass.: Allyn and Bacon, 1989.

Brosnac, D., *Guitar Electronics for Musicians*, New York: Amsco Publications, 1983.

Brown, M., *Power Supply Cookbook*, Boston: Butterworth–Heinemann, 1994.

Encyclopedia Britannica Inc., "Electromagnetism"; "Electromagnetic Radiation"; "Radio," *Encyclopedia Britannica*, Chicago: Encyclopedia Britannica Inc., 1999.

Fair-Rite Products Corp., *Choosing a Ferrite for the Suppression of EMI*, Fair-Rite Products Corp., 1992.

Feynman, R. P., R. B. Leighton, and M. Sands, *The Feynman Lectures on Physics Vol II: Mainly Electromagnetism and Matter*, Reading, Mass.: Addison-Wesley Publishing, 1964.

Flanagan, W. M., *Handbook of Transformer Design and Applications*, 2nd Edition, New York: McGraw-Hill, 1992.

Griffiths, D. J., *Introduction to Electrodynamics*, 3rd Edition, Upper Saddle River, NJ: Prentice Hall, 1999.

Grossner, N. R., *Transformers for Electronic Circuits*, New York: McGraw-Hill Book Company, 1967.

Halliday, D., R. Resnick, and J. Walker, *Fundamentals of Physics*, 2nd Edition, New York: John Wiley & Sons, 1986.

Hatsopoulos, G. N., and J. H. Keenan, *Principles of General Thermodynamics*, Malabar, FL: 1981.

Kraus, J. D., and D. A. Fleisch, *Electromagnetics with Applications*, 5th Edition, Boston: McGraw-Hill, 1999.

Paul, C. R., *Introduction to Electromagnetic Compatibility*, New York: John Wiley & Sons, 1992.

Shrader, R. L., *Electronic Communication*, 5th Edition, New York: McGraw-Hill, 1985.

Web Resources

Application notes on ferrites
http://www.fair-rite.com/

Technical articles concerning guitar pickups
http://www.till.com/articles/index.html

5 RADIATION

Every person, whether technical or nontechnical, is quite familiar with electromagnetic radiation. We directly experience radiation as light and heat radiation, and we also rely on radio, TV, cell phones, and other devices that make use of this property of nature. But what is electromagnetic radiation and how is it produced? How is radiation different from the static fields we find around a charged object or permanent magnet? Intuitively we know that radiating fields are far-reaching, whereas nonradiating fields like those of a refrigerator magnet are concentrated near their source. This intuitive notion is a good start for learning the details.

Understanding antennas and electromagnetic radiation is obviously important in RF engineering, in which capturing and propagating waves are primary objectives. An understanding of radiation is also important for dealing with the electromagnetic compatibility (EMC) aspects of electronic products, including digital systems. EMC design is concerned with preventing circuits from producing inadvertent electromagnetic radiation and stray electromagnetic fields. EMC also involves preventing circuits from misbehaving as a result of ambient radio waves and fields. With the ever-increasing frequencies and edge rates of digital systems, EMC becomes increasingly harder to achieve. The seemingly mystical process by which circuits radiate energy is actually quite simple.

STORAGE FIELDS VERSUS RADIATION FIELDS

First, it is important to define the terms involved. Unfortunately, there isn't a lot of consistency in terminology when discussing this subject. I will therefore explicitly define some terms. Electromagnetic fields can be divided into two basic types, storage fields and radiating fields. The main distinction between the two is that storage fields store energy in the vicinity of a source and radiating fields propagate energy away through free space. Storage fields can only exist in the vicinity (within several wavelengths) of conducting structures. Furthermore, storage fields disappear when their source is turned off. In contrast, radiating fields propagate forever—even after the source is turned off, radiation fields continue to exist.

For example, the famous astronomer Carl Sagan noted that if aliens were to point an antenna at Earth they would receive all our radio and TV broadcasts. Some of the energy of these signals escapes from the Earth and propagates out through space. One day, an alien astronomer in some distant part of the galaxy may point a radio telescope toward our solar system and pick up the broadcast of the first Super Bowl or an episode of *Baywatch*. Conversely, scientists on Earth are actively searching the skies for signs of any "intelligent" broadcasts not of earthly origin. The project is called SETI, the Search for ExtraTerrestrial Life. In contrast to earthly radiating fields, no alien observer has a chance of detecting the storage field of an earthly permanent magnet, even those magnets strong enough to pick up a car.

Another distinctive characteristic of radiating fields is that they must take the form of a wave, and they must have both electric and magnetic components. Furthermore, the electric and magnetic fields must be at right angles to each other. The direction of propagation is defined by $E \times B$. In other words, the direction of propagation is orthogonal to E and B, and can be determined by the cross product right hand rule given in Chapter 3. Moreover, in free space, the ratio of the magnitude of the electric field to that of the H-field is always $120\pi \cong 377$ ohms— the impedance of free space. The ratio of the electric field (E) to the magnetic field (B) is equal to the speed of light, $c \cong 3 \times 10^8$ m/s.

Storage fields can consist of waves, but they can also be static (DC) like a static charge. They can be exclusively electric or exclusively magnetic or any combination of the two. Within the umbrella of storage fields, I include the fields that surround electronic circuits that serve to store and transport energy in the circuit. For any powered circuit, the space surrounding every wire, resistor, capacitor, inductor, and transistor is permeated with some combination of electric and magnetic fields. I could be more precise and call these fields "storage/transport fields," but I think this term would be too cumbersome.

A commonly used term for the storage field is *near field*. This term is appropriate because the storage field is always concentrated near the source. The energy density contained in storage fields always decays at a rate quicker than $1/r^2$, where r is the distance to the source. In contrast, for radiating fields, the energy density decays at a rate of exactly $1/r^2$. Mathematically, this statement implies that the radiating field spreads out on the surface of a spherical shell of radius r. Because the surface area of a sphere is $4\pi r^2$, the total energy on any sphere is the same at any distance. For this reason, the radiating field is often called the *far field*. Unfortunately, the term far field is also sometimes used to describe the very far portion of the radiating field, where the radiation can be approximated as plane waves. For example, the sun's light rays can usually be considered as parallel (i.e., a plane wave) here on Earth because we are so far away from the sun.

The last major distinction between storage and radiating fields is how the source reacts when an observer absorbs some of the energy in the field. Consider a television broadcast. The power required for the TV station is independent of how many people have TVs and who has them turned on. If you buy a second TV for your home, the TV station does not need to turn up the power on their transmitting antenna. Once the signal leaves the vicinity of the antenna, it is gone forever, regardless of whether the power of the signal gets absorbed by a TV set, tree leaves, or rain, or is reflected into space. By tuning into a TV signal, you have no effect on the source that sent it. The same cannot be said of storage fields. Any time you try to measure or receive a signal in the near field, you will cause an effect on the source circuit. By extracting or diverting the energy of the near field, you are necessarily causing a reaction in the source circuit. For this reason the storage field is often called a *reactive field*. (Don't confuse the term reactive field with the terms reactive power and reactive impedance. These terms are all related but not in a one-to-one manner. For instance, a reactive field can be associated with real power transport.) Engineers familiar with EMC (electromagnetic compatibility) specification know how hard it is to make near field measurements; the very act of measuring changes the field you are trying to measure.

I will use the terms *storage field*, *near field*, and *reactive field* interchangeably.

ELECTRICAL LENGTH

The concept of electrical length (physical length divided by wavelength) is crucial to the understanding of antennas and other radiating systems. The power radiated from an antenna scales in terms of electrical length. In other words, two antennas with the same shape and electrical length will produce the same amount of radiation for a given current. For example, consider three antennas, each driven by a 1 Ampere signal and each designed for and driven at a different frequency: 60 Hz, 100 MHz, and 10 GHz. Each antenna will have the same radiation pattern and radiate the same amount of energy if they have the same geometry with equal dimensions as measured by electrical length. Such an antenna has a length of 2.5×10^6m, 1.5 m, and 1.5 cm for signals at 60 Hz, 100 MHz, and 10 GHz, respectively. From these numbers, it is quite obvious why practical radio systems use higher frequencies.

The Engineer and the Lawyer

A practical example will help to clarify these concepts. Consider the following fictitious disagreement. An electrical engineer is telling his lawyer friend about his latest home electronics project. The engineer lives near some high-voltage power lines and is working on a device that will harness the power of the 60 Hz electromagnetic field that permeates his property. The lawyer immediately states that what the engineer plans to do would, in effect, be stealing from the utility company. This statement angers the engineer who replies, "That's the trouble with you lawyers. You defend laws without regard to the truth. Even without my device, the stray electromagnetic energy from the power lines is radiated away and lost, so I might as well use it." The lawyer stands his ground and says the engineer will still be stealing.

Who is right? The lawyer is correct even though he probably doesn't know the difference between reactive and radiating electromagnetic fields. The field surrounding the power lines is a reactive field, meaning that it stores energy as opposed to radiating energy, so the engineer's device would in fact be "stealing" energy from the power lines. But why? Why do some circuits produce fields that only store energy, while others produce fields that radiate energy?

Circuits That Store and Transport Energy

To further examine this situation, consider the circuit in Figure 5.1A. It is a simple circuit consisting of an AC power source driving an inductor. If the inductor is ideal, no energy is lost from the power supply. The inductor does, however, produce an electromagnetic field. Because no energy is lost, this field is purely a storage field. The circuit pumps power into the field and at the same time the field returns power to the circuit. Because of this energy cycling, the current and voltage of the inductor are out of phase by 90 degrees, thus producing a reactive impedance, $Z_L = +j\omega L$.

Referring to Figure 5.1B, when a second circuit consisting of an inductor and resistor is placed near the first circuit, the field from L1 couples to L2 and causes current to flow in the resistor. The coupled fields of the two inductors create a transformer. Here the reactive field allows energy

Figure 5.1 A) An inductor creates a reactive field that stores energy. B) Adding a second inductor harnesses the reactive field to transfer energy to a load without metallic contact.

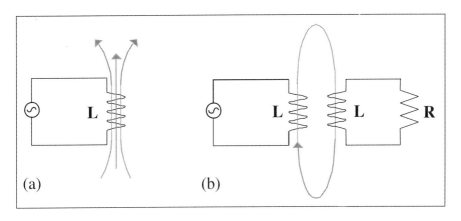

Figure 5.2 A) A capacitor creates a field that stores energy. B) Adding a second capacitor harnesses some of the storage field to transfer energy to the load resistor.

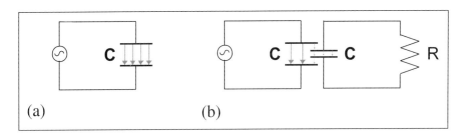

to be transferred from the source to the resistor even though the original circuit has not changed. This action suggests that a reactive field can store energy or transfer energy, depending on what other electrical or magnetic devices are in the field. So the reactive field *reacts* with devices that are present within it. In a reciprocal manner, a capacitor creates a reactive field that can store energy, transfer energy, or do both (Figure 5.2).

Circuits That Radiate

Now consider the circuits in Figure 5.3. Here an AC voltage source drives two types of ideal antennas, a half-wavelength loop and a half-wavelength dipole. Unlike the previous circuits, the antennas launch

Figure 5.3 The two most basic antennas are: A) a loop antenna whose circumference is equal to the source wavelength divided by two, and B) a dipole antenna whose length is equal to the source wavelength divided by two.

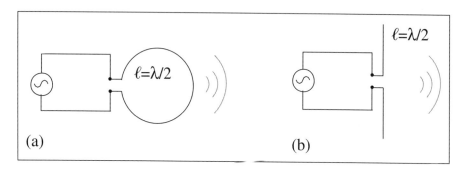

propagating fields that continuously carry energy away from the source. The energy is not stored, but propagates from the source regardless of whether there is a receiving antenna. This energy loss appears as resistance to the source, like the way loss in a resistor corresponds to heat loss.

Explaining the Lawyer's Claim

Now back to the story of the engineer and the lawyer. The engineer thought the power transmission line near his house was radiating energy like an antenna, and that he was just collecting the radiating energy with a receiving antenna. However, when the engineer measured the field on his property, he was measuring the reactive field surrounding the power transmission lines. When he activates his invention, he is coupling to the reactive field and removing energy stored in the field surrounding the power lines. The circuit he forms is analogous to the transformer circuit in Figure 5.1B, so the engineer is, in fact, stealing the power.

These examples illuminate the different characteristics of reactive and radiating electromagnetic fields, but they still do not answer the question of why or how radiation occurs. To understand radiation, it is best to start with the analysis of the field of a point charge.

THE FIELD OF A STATIC CHARGE

For a single charged particle, such as an electron, the electric field forms a simple radial pattern as shown in Figure 5.4. By convention, the field

THE FIELD OF A STATIC CHARGE

Figure 5.4 The static field of a positive charge is shown as a vector plot (top left), streamline plot (middle left), and a contour plot (bottom left). The static field of a dipole is shown as a vector plot (top right), streamline plot (middle right), and contour plot (bottom right).

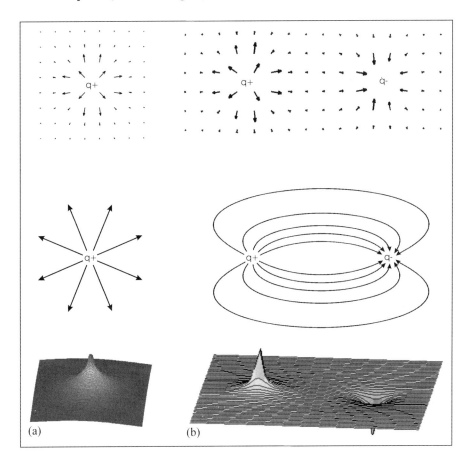

lines point outward for a positive (+) charge and inward for a negative (−) charge. The field remains the same over time; hence, it is called a static field. The field stores the electromagnetic energy of the particle. When another charge is present, the field exerts a force on the other object and energy is transferred. When no other charged particles are present, the field has no effect but to store energy. The fact that energy is transferred from the field only when another charged particle is present is a defining characteristic of the static field. As you will soon learn, this fact does not hold true for a radiating field.

THE FIELD OF A MOVING CHARGE

Now consider the same charged particle moving at a constant velocity, much lower than the speed of light. The particle carries the field wherever it goes and at any instant the field appears the same as a static field (see Figure 5.5A). In addition, because the charge is now moving, a magnetic field also surrounds the charge in a cylindrical manner as governed by the Biot-Savart law. As in the case of a static charge, both the electric and magnetic field of a constant velocity charge serve to store energy and transmit the electric and magnetic forces only when other charges are present. To make the description easier, I'll ignore the magnetic field for most of the chapter.

THE FIELD OF AN ACCELERATING CHARGE

When a charged particle is accelerated, something interesting occurs. The lines of the electric field start to look bent, as shown in Figure 5.5B. A review of Einstein's theory of relativity helps to explain why bending occurs: No particle, energy, or information can travel faster than the speed of light, c. This speed limit holds for fields as well as particles. (For that matter, quantum electrodynamics says that a field is just a group of virtual particles called "virtual photons," as described in Chapter 6.) For instance, if a charged particle were suddenly created, its field would not instantly appear everywhere. On the contrary, the field would first appear in the immediate region surrounding the particle and then extend outward at the speed of light. For example, light takes about eight minutes to travel from the sun to the Earth. If the sun were to suddenly extinguish, we would not know until eight minutes later. Similarly, as a particle moves, the surrounding field continually updates to its new position, but this information can propagate only at the speed of light. Points in the space surrounding the particle actually experience the field corresponding to where the particle was at a previous time. This delay is referred to as *time retardation*. It seems reasonable to assume that even a charge moving at constant velocity should cause the field lines to bend due to this time retardation. However, Einstein's theory of relatively states that velocity is a relative measurement, not an absolute measurement. If the field lines are straight in one inertial observer's reference frame, they must be straight in all other inertial observer's reference frames. The discussion of relativity in Chapter 6 will clarify this situation. For now, you must take the claim on faith.

THE FIELD OF AN ACCELERATING CHARGE

Figure 5.5 The plots show the electric field for a charge in several different states of motion. Particle locations and field lines at earlier times appear in gray.

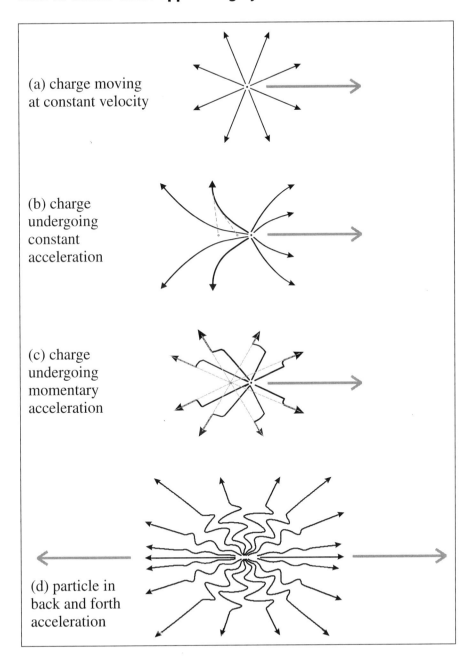

(a) charge moving at constant velocity

(b) charge undergoing constant acceleration

(c) charge undergoing momentary acceleration

(d) particle in back and forth acceleration

A Curious Kink

To understand why bent field lines of a charge correspond to radiated energy, consider a charged particle that starts at rest and is "kicked" into motion by an impulsive force. When the particle is accelerated, a kink appears in the field immediately surrounding the particle. This kink propagates away from the charge, updating the rest of the field that has lagged behind, as shown in Figure 5.5C. Part of the energy expended by the driving force is expended to propagate the kink in the field. Therefore, the kink carries with it energy that is electromagnetic radiation. Fourier analysis shows that since the kink is a transient, it will consist of a superposition of many frequencies. Therefore, a charge accelerated in this manner will radiate energy at many simultaneous frequencies.

Keep in mind here that I am discussing the field of an individual charge. I don't mean to imply that bent field lines always imply radiation. Figure 5.4B shows a simple example of two charges producing bent field lines. No radiation occurs in this static setup. It is the acceleration that produces the radiation.

X-RAY MACHINES

A charge that is decelerated will also radiate energy because deceleration requires a force and corresponding energy. For example, when a moving charge is brought to a halt, the charge will radiate at many simultaneous frequencies as described by Fourier analysis. If a high-speed charge is abruptly decelerated, perhaps by a collision with a much heavier object, a large amount of energy will be radiated. The X-rays of medical diagnostic equipment are produced in this very manner. Electrons are brought up to high speeds by placing them in a very large electric field. The high-speed electrons are aimed at a massive metallic target, where they are abruptly stopped and absorbed. During the violent collision, high-energy photons are radiated. Most of these photons are in the X-ray band.

THE UNIVERSAL ORIGIN OF RADIATION

Radiation from charges can also be analyzed from a kinetic energy perspective. Newton's second law states that a force is needed to accelerate a particle. The force transfers energy to the particle, thus increasing its kinetic energy. The same analysis holds true for the particle's field. Energy is required to accelerate the field. This energy propagates outward as a wave (see Figure 5.6A), increasing the kinetic energy of the field.

All electromagnetic radiation, be it RF radiation, thermal radiation, or optical radiation, is created this way—by changing the energy of electrons or other charged particles. This general statement includes not only changes in energy of free electrons due to acceleration/deceleration, but also the change in energy of electrons bound in atoms due to change in orbitals (quantum energy state changes).

The fact that the field requires a force and corresponding energy for it to be accelerated implies that the field has inertia or mass. I will explore this topic in depth in Chapter 6.

THE FIELD OF AN OSCILLATING CHARGE

A charge moving in a circle experiences sinusoidal acceleration. In fact, sinusoidal acceleration will occur for a charge moving in any oscillatory manner. In this case, the "kinks" in the field will be continuously varying and sinusoidal and the electromagnetic radiation will occur only at the frequency of oscillation. As shown in Figure 5.6B, an oscillating charge produces a rippling of waves that propagate outward, in some ways similar to the water waves produced by tossing a rock into a pool of water.

Another analogy will help illustrate the creation of waves from a charge. Imagine a ball with long flexible spikes or rods sticking out from its surface, like a porcupine. Here, the spikes represent the electric field lines. When the ball is still, the rods stick straight out. Furthermore, when the ball is in uniform straight-line motion, the rods will also stick straight out (assuming that you ignore the effects of air friction or that the ball is moving in a vacuum). However, if you shake the ball back and forth, a sinusoidal wave will propagate outward along the flexible rods, like a wave along a rope. (Figure 5.5D helps to illustrate this concept.)

THE FIELD OF A DIRECT CURRENT

If a constant voltage is connected across a length of wire, the voltage will cause a proportional current governed by Ohm's law ($I = V/R$). The

Figure 5.6 These two sequences show the electric field of accelerated charges. A) A charge starts at rest and is accelerated by a short impulsive force. B) A charge starts at rest and is sinusoidally accelerated along the horizontal axis.

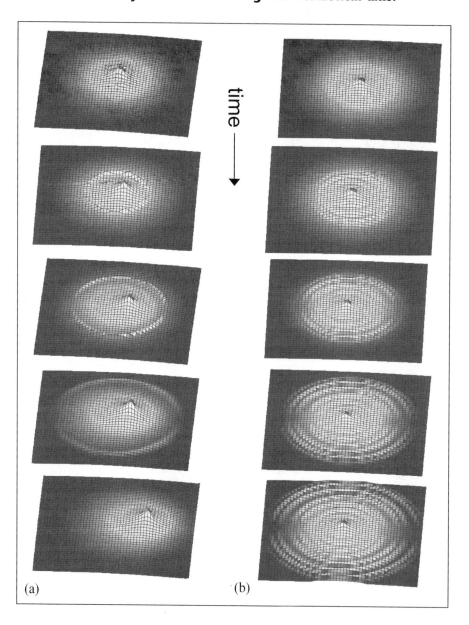

THE FIELD OF A DIRECT CURRENT

Figure 5.7 Electric field surrounding a wire carrying a DC current. Shades of gray denote the relative voltage levels inside the wire. Arrows denote the current.

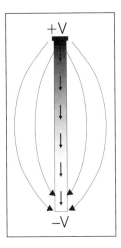

DC current traveling in a wire consists of the migration of electrons. Although the path of each individual electron is quite random and complex, the average movement of the electrons, considered as a group, produces a constant drift of charge. Therefore, at a macroscopic level we can use an equivalent model that consists of fictitious charges traveling at a constant velocity, ignoring the specifics of each electron. Radiation will not occur, because the effective charge is traveling at a constant velocity and experiences no acceleration. Figure 5.7 illustrates the field of a DC current.

The randomness of the electron movement is caused by collisions at the atomic level. These fluctuations show up as noise in the circuit. These noisy variations of the individual electrons do in fact cause radiation, but due to the violent nature of the electron collisions, most of the radiation energy occurs in the infrared region, and is commonly known as heat radiation. As the current is increased, more collisions occur, resulting in a hotter wire and more heat radiation. Some of this radiation is propagated at lower energies, in the microwave and radio bands. The radio wave radiation shows up as white noise when received by an antenna because of its random nature. For example, when you tune your radio to a frequency where no broadcast is present, all you hear is the noisy waves of thermal radiation.

Figure 5.8 Electric field surrounding a wire carrying a slowly varying AC current. A) Electric field at time $t = 0$. B) Electric field at time $t = T/2$, a half-cycle later.

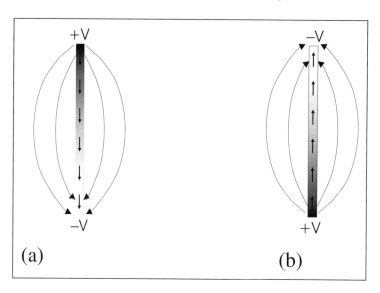

THE FIELD OF AN ALTERNATING CURRENT

If we allow the voltage source across a wire to slowly oscillate in time at frequency, f_o, the accompanying electric field will take the same form as that of the DC charge, except that the magnitude will vary between positive and negative values (see Figure 5.8).

Relating frequency to wavelength by $\lambda = c/f$, we define "slow oscillation" as any frequency whose corresponding wavelength is much greater than the length of the wire. This condition is often called *quasi-static*. In this case, the current in the wire will vary sinusoidally, and the effective charge will experience a sinusoidal acceleration. Consequently, the oscillating charge will radiate electromagnetic energy at frequency, f_o. The power (energy per time) radiated is proportional to the magnitude of current squared and the length of the wire squared, because both parameters increase the amount of moving charge. The radiation power is also proportional to the frequency squared since the charge experiences a greater acceleration at higher frequencies. (Imagine yourself on a spinning ride at an amusement park. The faster it spins, the greater the acceleration you and your lunch feel.)

Expressed algebraically:

Figure 5.9 Electric field surrounding a wire carrying a rapidly varying AC current. A) Electric field at time $t = 0$. B) Electric field at time $t = T/2$, a half-cycle later.

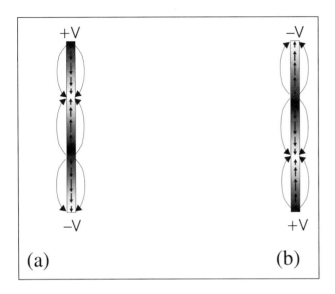

$$\sqrt{\text{Radiated power}} \sim \text{current} \times \text{length} \times \text{frequency}$$

This expression shows clearly why RF signals radiate more readily than signals at lower frequencies such as those in the audio range. In other words, a given circuit will radiate more at higher frequencies. Because wavelength is inversely proportional to frequency ($\lambda = c/f$), an equivalent expression is

$$\sqrt{\text{Radiated power}} \sim (\text{current}) \times (\text{electrical length})$$

Hence, at a given source voltage and frequency, the radiated power is proportional to the length of the wire squared. In other words, the longer you make an antenna, the more it will radiate.

Up until now we have considered only slowly oscillating fields. When the frequency of the voltage source is increased such that the wavelength approaches the length of the wire or less, the quasi-static picture no longer holds true. As shown in Figure 5.9, the current is no longer equal throughout the length of wire. In fact, the current is pointing in different directions at different locations. These opposing

Figure 5.10 The root-mean-square radiation power versus electrical length for a dipole antenna driven by 1 ampRMS of current. The radiation resistance follows the same curve. At one-half wavelength the resistance is 73 ohms.

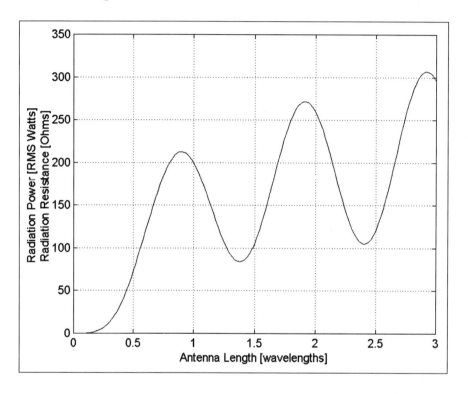

currents cause destructive interference just as water waves colliding from opposite directions tend to cancel each other out. The result is that the radiation is no longer directly proportional to the wire or antenna length squared.

Figure 5.10 shows a plot of radiated power as a function of antenna length. When the antenna is much smaller than a wavelength, the radiated power is proportional to the length squared. However, for wire lengths near or above the wavelength, the radiated power relates as a slowly increasing and oscillating function. This "diminishing returns" of the radiation power versus wire length is one of the reasons that $\lambda/2$ is usually chosen as the length for a dipole antenna ($\lambda/4$ for a monopole). The other reasons being that at $\ell = \lambda/2$, the electrical impedance of the antenna is purely real (electrically resonant), and the radiation pattern is simple (single lobed) and broad.

NEAR AND FAR FIELD

As mentioned earlier, an AC circuit will have a reactive field and a radiating field. The reactive field of an AC source circuit or system is often referred to as the near field because it is concentrated near the source. Similarly, the radiating field is referred to as the far field because its effects extend far from the source. Let's examine why.

The power density of an electromagnetic field at a distance, r, from the source can be represented by a series in $1/r$,

$$\text{Field power density} = P_d = C_1/r^2 + C_2/r^3 + C_3/r^4 + \ldots$$

Now, imagine a sphere with radius r centered at the source. The total power passing through the surface of the sphere can be calculated by multiplying the power density by the surface area of the sphere,

$$\text{Total power leaving sphere} = P = (4\pi r^2)P_d = 4\pi(C_1 + C_2/r + C_3/r^2 \ldots)$$

Examining this formula, you can see that the first term is purely a constant. For this term, no matter what size you make the sphere, the same amount of power is flowing through it. This result is just a mathematical way of showing that power is carried away from the source. Therefore, the first term is due solely to the radiated field. Another thing to notice is that as r gets very large all the other terms will become negligible, leaving only the radiated term. For this reason the radiated field is often called the far field. Conversely, at very close distances (small values of r), the nonconstant terms will become much larger and the constant radiating term will become negligible. These nonconstant terms taken together represent the power in the reactive field, and since it dominates at close distances it is called the near field.

For a dipole antenna, the near-field and far-field boundary is generally considered to be about $\lambda/(2\pi)$. Furthermore, the reactive field typically becomes negligible at distances of 3λ to 10λ for a dipole. It is interesting to compute the boundary at different frequencies. At 60 Hz, the boundary is 833 km. Therefore, almost all cases of 60 Hz interference occur in the near (reactive) field. At 100 MHz, the boundary is 1/2 meter, making this frequency useful for radio communication. At 5×10^{14} Hz (optical waves), the boundary is 0.1 µm, explaining why optical sources such as light bulbs always appear as radiating sources and never as reactive sources.

The near field and far field have other characteristics. The shape of the near field is very closely related to the structure of the source,

Figure 5.11 Compare the wave impedance as a function of distance given for a loop antenna (as in Figure 5.3A) and a dipole (as in Figure 5.3B). In the near field, the loop antenna field energy is mostly magnetic but the dipole antenna field energy is mostly electric. In the far field, the energy partitioning is the same for both antenna types.

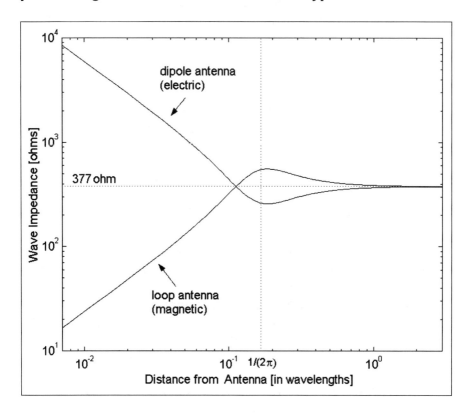

whereas the far field becomes independent of the source, taking the form of spherical waves. At very large distances, the far field takes the form of traveling plane waves. The specific requirement for the plane wave approximation is $r > 2(d_s + d_r)^2/\lambda$, where d_s is the size of the source antenna, d_r is the size of the receiving antenna, and r is the distance between the antennas. The wave impedance (ratio of electric to magnetic field magnitude) of the near field is also a function of the source circuit, whereas in the far field the wave impedance is dependent only on the medium ($\eta \cong 377\,\Omega$ in free space). Figure 5.11 graphs the wave impedance as a function of distance.

The field characteristics are summarized in the following table.

	Near (Reactive) Field	Far (Radiated) Field
Carrier of Force	Virtual photon	Photon
Energy	Stores energy. Can transfer energy via inductive or capacitive coupling.	Propagates (radiates) energy.
Longevity	Extinguishes when source power is turned off.	Propagates until absorbed.
Interaction	Act of measuring field or receiving power from field causes changes in voltages/currents in source circuit.	Act of measuring field or receiving power from field has no effect on source.
Shape of Field	Completely dependent on source circuit.	Spherical waves. At very long distances, field takes shape of plane waves.
Wave impedance	Depends on source circuit and medium.	Depends solely on propagation medium ($\eta = 120\pi \cong 377\,\Omega$ in free space).
Guiding	Energy can be transported and guided using a transmission line.	Energy can be transported and guided using a wave guide.

THE FRAUNHOFER AND FRESNEL ZONES

To be even more precise, each electromagnetic field can be divided into four zones: the near zone, the intermediate zone, the far zone, and the plane-wave zone. The near zone is the portion of the field close to the source. It is defined as the region where stored energy is much greater than any radiating energy. The far zone is the region where: 1) the stored field energy is much less than the radiating energy, 2) the wave impedance is approximately $\eta_o = 120\pi$, and 3) the electric and magnetic fields are perpendicular to one another. The intermediate zone is the region between the near and far zones. The plane-wave zone is the region in the far zone where the radiation can be approximated as plane waves. This last zone is different from the others because its definition depends on the size of the receiving antenna. Just as a basketball's surface appears curved to a human but relatively flat to a tiny microorganism on the surface, and the Earth's surface appears flat to a walking human, the apparent flatness of a curved wavefront depends on the relative size of the observer. In optics, the plane-wave zone is called the *Fraunhofer zone*, and the combination of the three other zones is called the *Fresnel zone*.

PARTING WORDS

In summary, stationary charges and charges moving with constant velocity produce storage fields (reactive fields); accelerated charges produce radiating fields in addition to the reactive field. DC sources cause a constant drift of charges and hence produce reactive fields. AC sources cause the acceleration of charges and produce both reactive and radiating fields. Radiating fields carry energy away from the source regardless of whether there is a receiving circuit or antenna. In the absence of another circuit, reactive fields store energy capacitively and/or inductively. In the presence of another circuit, reactive fields can transfer energy through inductive or capacitive coupling. In general, radiation increases with frequency and antenna length. Similarly, radiation (as well as transmission line effects) is usually negligible when wires are much shorter than a wavelength. The reactive field is also referred to as the near field, and its characteristics are very dependent on the source circuit. The radiating field is referred to as the far field, and many of its characteristics, such as wave impedance, are independent of the source.

Finally, I indirectly compared a 60 Hz transmission line to an antenna. In reality, a uniform transmission line is typically a poor radiator at any frequency because of its geometry. For example, the wires of a two-wire transmission line have opposite currents and charge. Hence, at far distances the fields of each wire tend to cancel each other. Thus, even the power grid transmission lines that span hundreds of miles in the southwest United States do not radiate much energy. By using uniform transmission lines, the power companies lose hardly any power to radiation even when the lines are electrically long.

BIBLIOGRAPHY

Bansal, R., "The Far-field: How Far Is Far Enough?" *Applied Microwave & Wireless*, November, 1999.

Epstein, L. C., *Thinking Physics—Is Gedanken Physics; Practical Lessons in Critical Thinking*, 2nd Edition, San Francisco: Insight Press, 1989.

Feynman, R. P., R. B. Leighton, and M. Sands, *The Feynman Lectures on Physics Vol I: Mainly Mechanics, Radiation, and Heat*, Reading, Mass.: Addison-Wesley Publishing, 1963.

Griffiths, D. J., *Introduction to Electrodynamics*, 3rd Edition, Upper Saddle River, NJ: Prentice Hall, 1999.

Heald, M., and J. Marion, *Classical Electromagnetic Radiation*, 3rd Edition, Fort Worth, Texas: Saunders College Publishing, 1980.

King, R. W. P., and T. T. Wu, "The complete electromagnetic field of a three-phase transmission line over the earth and its interaction with the human body," *Journal of Applied Physics*, Vol. 78, No. 2, July 15, 1995.

Kraus, J. D., *Antennas*, 2nd Edition, Boston: McGraw-Hill, 1988.

Kraus, J. D., and D. A. Fleisch, *Electromagnetics with Applications*, 5th Edition, Boston: McGraw-Hill, 1999.

Lo, Y. T., and S. W. Lee, Editors, *Antenna Handbook—Theory, Applications, and Design*, New York: Van Nostrand Reinhold Company, 1988.

Omar, A., and H. Trzaska, "How Far Is Far Enough," *Applied Microwave & Wireless*, March 2000.

Paul, C. R., *Introduction to Electromagnetic Compatibility*, New York: John Wiley & Sons, 1992.

Purcell, E. M., *Electricity and Magnetism*, 2nd Edition, Boston: McGraw-Hill, 1985.

Schmitt, R., "Understanding Electromagnetic Fields and Antenna Radiation Takes (almost) No Math," *EDN*, March 2, 2000.

Slater, D., *Near-Field Antenna Measurements*, Boston: Artech House, 1991.

Web Resources

Some great animated tutorials on radiation can be found at:
http://www.Colorado.EDU/physics/2000/waves_particles/index.html
http://www.Colorado.EDU/physics/2000/xray/index.html
http://webphysics.ph.msstate.edu/jc/library/20-9/index.html
http://www.amanogawa.com/Virtuals/Amanogawa/waves.html
http://hibp.ecse.rpi.edu/~crowley/javamain.htm
http://ostc.physics.uiowa.edu/~wkchan/EM/PROGRAMS/POLARIZATION/

6 RELATIVITY AND QUANTUM PHYSICS

Relativity and quantum physics may seem out of place in a book on electromagnetics, but each is crucial to the foundations of how electromagnetic fields operate. To a large extent this chapter is an excursion from what is meant to be a practical handbook, but I think the topic is too interesting not to cover. Both relativity and quantum physics are quite fascinating and each will give you a whole new perspective on electromagnetics. Unbeknownst to many people, relativity is at work behind the scenes of many electromagnetic phenomena. Quantum physics may not be necessary for most applications, but it is crucial for understanding lasers, the basis of fiber optics. The photon is known by every technical person as a packet of light energy. What may not be known is that all electromagnetic energy consists of photons. Photons are exclusively the realm of quantum physics.

RELATIVITY AND MAXWELL'S EQUATIONS

You may be wondering how electromagnetics changes when special relativity is included. Or to put the question in historical context: How were Maxwell's equations modified to conform to Einstein's relativity? It is well known that Newton's basic tenets of physics and the concept of space and time developed mainly by Galileo and Newton had to be radically changed to accommodate relativity. You may be surprised to learn that there were no changes made to Maxwell's equations!* In fact, special relativity follows directly from Maxwell's equations.

I will now return to the story of Maxwell's equations and the aether. Maxwell's equations predict that electromagnetic waves exist and that their speed is $\sim 3 \times 10^8$ m/s, the speed of light. However, the equations

*Note that Maxwell's equations must be used in conjunction with the laws of relativistic mechanics, as opposed to Newtonian mechanics, to be in accord with special relativity. Thus, charge density and current density must be multiplied by the relativistic stretch factor, γ.

do not state what this speed is relative to. For example, when a car driving down the highway turns on its lights, the light travels away at the speed of light, relative to the car. The same is true for the radio waves sent out by a cellphone in the car. Assume the car is traveling at 60 miles per hour. Ordinary experience tells us that if you are standing still on the side of the highway, you would measure the speed of both the light waves from the headlights and the radio waves from the cellphone to be the speed of light plus 60 miles per hour. Maxwell's equations, however, state that the speed of radio waves and light waves is always the same under any measurement. How could the driver and you measure the speed of the light from the headlights to be the same? This question bothered the physicists of the late 19th century and early 20th century. Most physicists believed for this reason that Maxwell's equations were flawed and needed to be modified to match the prevailing view of space and time.

Einstein was an imaginitive thinker. He said that Maxwell's equations are correct as they were first written, and that it was Newton's laws of physics and the Newton-Galileo laws of space and time that needed to be modified. Obviously Einstein made a bold statement.* He also had the physics to back up his statement. It took decades, but Einstein's theory of relativity gradually gained full acceptance of the scientific community after countless experiments proved him to be correct. At this point in the 21st century, I think it is safe to say that Einstein was correct and perhaps his theory should now be called the law of relativity. Close to 100 years of scrutiny and experimentation have passed without a single failure. Of course, general relativity isn't the end of physics. Most notably, relativity and quantum physics have yet to be merged. There has been hope in recent years that string theory will provide the unified theory of physics, commonly called the Theory of Everything.

I cannot in one chapter teach the entirety of relativity, but I can highlight the most important aspects of relativity and mention the relation of relativity to electromagnetics. The basic premise behind Einstein's work is that we should be able to write the laws of physics such that any event or phenomenon can be described by the same laws, regardless of who does the measurement. In other words, there should be a universal set of laws to describe observations, regardless of whether the observer is stationary, moving at a constant speed, or accelerating. More-

---

*In this abreviated history of relativity, I leave out the contributions of Michelson and Morley, Poincaré, Lorentz, and FitzGerald. Refer to the books by Blatt (1989), Feynman, Leighton, and Sands (1963), and Moore (1995) for a more detailed history.

over, there must be a set of equations to translate measurements from one reference frame to another. When Einstein first introduced his theory in 1905, he covered the topics of stationary and constant speed observers. This subset of relativity is called *special relativity*. Later, in 1916, Einstein introduced his theory of *general relativity*, which not only provided the framework for any type of motion, including acceleration, but also provided a new theory of gravity. Gravity is not like other forces. Gravity manifests itself as the curvature (warping) of space-time. Mass causes space-time to curve, and the curvature of space-time determines how other masses move.

For the understanding of electromagnetics, special relativity is all that is really needed. The entirety of special relativity can be summarized in two simple statements:

- Light travels at the same speed when measured by any observer.
- The laws of physics are the same in every inertial (gravitational and acceleration free) reference frame.

From these two simple statements, all of special relativity can be derived. The first statement is a consequence of Maxwell's equations, so in some sense it is really only the second equation that defines special relativity. A reference frame is the coordinate system (including the three dimensions of space and one dimension of time) of the observer who is performing the measurement. The term *inertial* basically means that the observer is not being accelerated. Accelerated motion is very different from uniform motion. Newton's first law states that a body in uniform motion will continue in uniform motion unless a force acts upon it. Newton's second law states that the mass of an object times the acceleration of the object is equal to the net force acting on the object. The force provides the energy to change the object's velocity. You learned the electromagnetic consequence of Newton's laws in Chapter 5. Electromagnetic radiation can be produced only if there is a force acting on a charge. This force provides the energy of the radiation. If I were to hold a charged ball in my hand and wave it back and forth, it would radiate energy at the frequency at which I wave it. I provide the energy that is radiated away. In other words, I must burn calories to move my arm, some more calories to move the ball, and even more calories to move the ball's electric field. Hence, the electric field itself has inertia or mass, just as the ball has mass.

There are further differences between uniform straight-line motion (inertial motion) and accelerated motion (noninertial motion). If you are sealed inside a windowless compartment and are traveling in uniform motion, there is no way for you to determine how fast you are going or whether you are moving or not. Uniform straight-line motion is completely relative. For instance, if you are traveling in uniform motion in a car or plane, the laws of physics are no different. You don't feel any different when you are in a plane traveling several hundred miles an hour. If you drop a ball, it falls straight down in your frame of reference. On the other hand, if the plane is accelerating, you can feel it. The movement of liquid in your inner ear is the main source for determining acceleration. Furthermore, if you drop a ball in an accelerating airplane or car, it doesn't travel straight down. It curves toward the back of the vehicle as it falls because once it leaves your hand it is no longer accelerating, whereas you and the vehicle still are accelerating.

The Speed of Light Is Always the Same, or "$c + v = c$"

Let's return to the story of the moving car on the highway. If the speed of the light waves and of the radio waves is the same for both the driver and you, something must account for this strange result. Mathematically we can express the speed of light as

$$c = \text{velocity of light} = \text{distance/time} = \Delta x / \Delta t$$

Indeed, for the light to be the same speed for all inertial observers, each must have a different reference of space and time. For the driver, the light travels away at a speed of $c \cong 3 \times 10^8 \text{m/s} = 300{,}000 \text{km/s}$. Therefore when one second elapses on the driver's watch, the light travels 300,000 km as measured by his meterstick. To make the mathematics simpler and the result more obvious, assume that the car is traveling at 100,000 km/s. Assuming that you do not know Einstein's relativity, you assume that the light's velocity will be the sum of the car velocity plus the speed at which the light left the car. This assumption seems reasonable. For example, if your friend is riding a bike at 10 miles/hour and throws a ball at 20 miles/hour, you (standing still) will measure the ball to move at 30 miles/hour. Thus, you predict that the light will have a velocity

$$300{,}000 \text{km/s} + 100{,}000 \text{km/s} = 400{,}000 \text{km/s}.$$

To test this hypothesis, you measure how far the light travels in one second. Your (incorrect) calculation shows that in one second the light should travel 400,000 km. To your amazement, the measurements show that the light has traveled only 300,000 km. This result implies that speeds do not add in the direct manner that you are probably accustomed to. The correct formula for adding the speed of two objects is:

$$v = \frac{v_1 + v_2}{1 + \frac{v_1 \cdot v_2}{c^2}} \approx v_1 + v_2$$

where the approximation on the right-hand side is only valid when both of the velocities are much less than the speed of light. I leave it to you to verify that if either velocity, v_1 or v_2, is equal to c, then the total velocity is also c. It may occur to you that the light emitted from the car's headlights should have more energy than light emitted from a stationary object. Even though the velocity is the same, the light emitted from the moving car does have more energy because its frequency is greater. This effect is the Doppler effect of light waves. The Doppler frequency shift of microwaves is used by police to measure the speed of passing cars.

Proper Length and Proper Time

So far, I have assumed that each observer (the driver and you) makes measurements using his/her own meterstick and stopwatch for distance and time measurements. I also assumed that the metersticks are exactly the same length when both are at rest and placed side by side. Physicists say that the proper length of each meterstick is exactly 1 meter. The term *proper length* is used to describe the length of an object when measured in the same reference frame as the object. In other words, the measurement is made with the observer and object at rest, relative to one another.

Furthermore, the stopwatches are synchronized and record time at the same rate when both are at rest. The term *proper time* is used to describe time measured by a clock traveling with the observer.

SPACE AND TIME ARE RELATIVE

To explain the constancy of the speed of light, relativity requires that space and time are different for observers who are moving relative to each other. There are three space-time effects of relativity: time dilation, time desynchronization, and length contraction. Figure 6.1 illustrates space and time as observed by you and how the space-time of the car driver maps into your space-time. For simplicity, only one spatial

116   RELATIVITY AND QUANTUM PHYSICS

Figure 6.1 Space-time map. Axes of the reference observer (you) are shown in black. Axes of the other observer (driver) are shown in gray. The driver's time axis has a slope, v/c. The driver's distance axis has a slope c/v. Both of the driver's axes are also stretched by a factor $\gamma \cdot \sqrt{1+(v/c)^2}$. Event A occurs at $t = 5$ sec, $x = 5$ light-secs, in your frame and at $t' = 3.5$ sec, $x' = 3.5$ light-secs, in the driver's frame. Event B occurs at $t = 7.1$ sec, $x = 7.1$ light-secs, in your frame and at $t' = 5$ sec, $x' = 5$ light-secs, in the driver's frame.

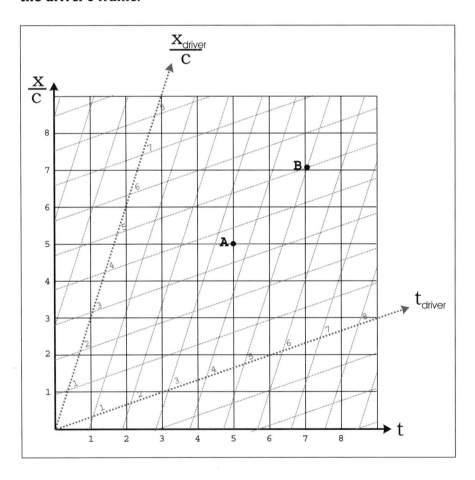

dimension is shown. The driver's space and time axes are tilted inward and stretched by the factor

$$\gamma \cdot \sqrt{1+\left(\frac{v}{c}\right)^2}$$

The variable γ is called the *stretch factor*. Its value is

$$\gamma = \frac{1}{\sqrt{1-\left(\frac{v}{c}\right)^2}}$$

where v is the velocity of the car with respect to you. When referring to the figures, assume that the car moves in the positive x direction. Notice that the x-axis is displayed in units of light-seconds (x/c), which is the distance that light travels in one second. The Lorentz transformation can be used to convert events from the driver's frame to your reference frame:

$$t = \gamma\left(t' + \frac{v}{c}x'\right)$$

$$x = \gamma\left(\frac{v}{c}t' + x'\right)$$

The Expansion of Time

Time changes for moving objects and observers. Time proceeds at a faster rate for you as compared to the driver. The formula for time dilation is $dt = \gamma \times d\tau$, where dt is the rate of the time for you, $d\tau$ is the rate of time for the driver, and γ is the stretch factor. If the car is traveling at one-third the speed of light, your time proceeds at a rate that is $dt/d\tau \cong 1.061$ times faster than the driver's. Keep in mind that neither of you feels any different, the difference in time is relative to each of you. Figure 6.2 illustrates this effect.

Simultaneity Is Relative

The loss of time synchronization is the strangest of the three space-time effects. Events that are simultaneous in one reference frame will not be simultaneous in a moving reference frame, and vice versa. For instance, suppose the car driver simultaneously turns on two lights. One light is at the front, and the other light is at the back of the car. In this situation, you do not observe the lights turning on simultaneously. You, in fact, observe the rear light turning on first and then $\Delta t = \gamma \times L \times v/c^2$ seconds later, the front light turns on. Figure 6.3 illustrates how this effect happens.

Figure 6.2 Time Dilation. Seven seconds in the drivers space-time corresponds to 7.42 seconds in your space-time.

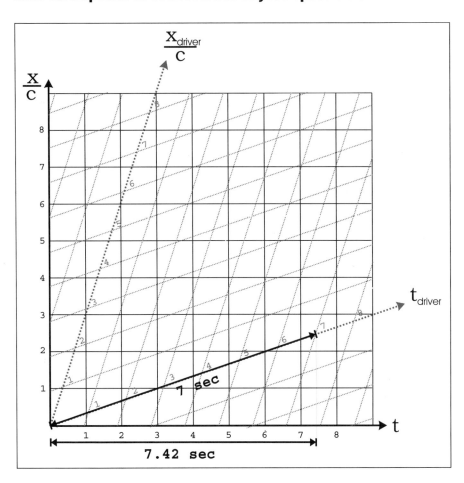

Lorentz Contraction of Length

Relativity states that your meterstick and the driver's meterstick appear to be of different lengths. Distance contracts in the direction of travel. The contraction of distances is known as Lorentz-FitzGerald contraction, named after the two scientists who independently derived the formulas using Maxwell's equations in the late 1800s. The contracted length of a moving object is

$$L' = \frac{L}{\gamma}$$

Figure 6.3 Loss of time synchronization. Two events, a light flash at $x' = 2$ light-sec and a light flash at $x' = 8$ light-sec, take place simultaneously in the driver's reference frame. In your reference frame, the flashes take place at separate times. The flash at the back of the car occurs about 2.1 seconds earlier than the other flash.

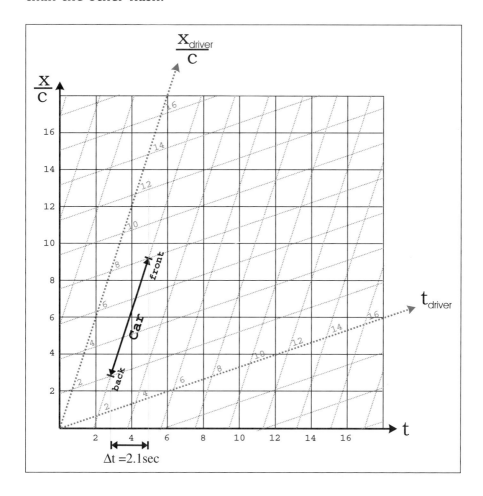

where L is proper length of the car, L' is the length as you measure it, and γ is the stretch factor. Therefore, if the proper length of the car is 6 light-seconds, and the car is traveling at one-third the speed of light, then the length of the car measured by you is $L' = 6/1.061 \approx 5.66$ light-seconds. At first it may seem that the length should expand rather than contract, in the same way that time expands. However, due to the desynchronization of time, the length of the car actually contracts. Let's say you measure the car by taking a photo of it using a flash camera. You can then calculate the length of the car from the photo, assuming that you know the details of the camera lens. You take the photo as the car zooms by, and from your reference frame, the light of the flash bulb hits the car simultaneously. However, due to the desynchronization of time, the driver sees the flash hit his car first in the front, and then the flash sweeps across the car. Therefore, by the time the light flash reaches the back of the car, it has moved forward a considerable amount. You measure a shorter car because you are really measuring the back end of the car at a time later than you measured the front of the car! Figure 6.4 illustrates this effect graphically. Figure 6.5 shows how light travels at the same speed in both reference frames.

SPACE AND TIME BECOME SPACE-TIME

As I have just shown, in relativity, space and time are not independent entities. On the contrary, space and time are intimately connected. The three dimensions of space and the one dimension of time form a four-dimensional geometry, called Minkowski geometry, named after Hermann Minkowski, who derived the formal mathematics in 1908. At the core of his work is his equation for calculating distances in relativity. The so-called *metric equation* is

$$\Delta s^2 = (c \cdot \Delta t)^2 - [\Delta x^2 + \Delta y^2 + \Delta z^2]$$

where Δs is called the space-time interval, Δt is the time difference between the events, and $[\Delta x, \Delta y, \Delta z]$ is the distance between two observed events. Notice that the spatial portion of this formula is just the Pythagorean theorem for determining distance in three dimensions. The beauty of this equation is that it holds true for any observers of any two events. In other words, the distances and times may be different for two observers, but the space-time interval, Δs, is always the same. *Many quantities are relative, but not everything is relative.*

THE COSMIC SPEED LIMIT AND PROPER VELOCITY

Another consequence of special relativity is that nothing can move faster than the speed of light. Any object with a non-zero rest mass requires infinite energy to reach the speed of light. Furthermore, rela-

Figure 6.4 Lorentz contraction. You measure the moving car at time $t = 6$ secs by taking a photograph. In the driver's frame, the parts of the car are photographed at different points in time. The front of the car is photographed at $t' = 2.8$ sec and the back of the car is photographed at $t' = 4.9$ sec. The rest of the car is photographed at intermediate times. This effect causes the car to appear shorter in your frame.

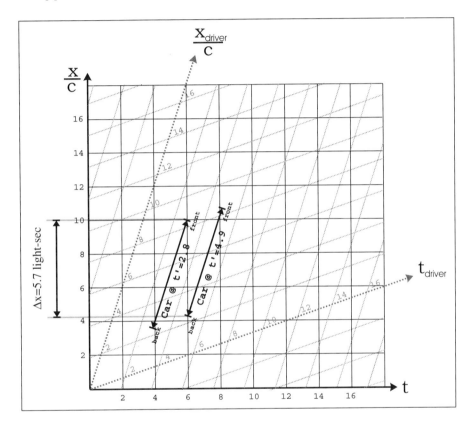

tivity implies that allowing objects to travel faster than the speed of light would also allow objects to travel backward in time.

Consider an astronaut who wants to travel to another star that is 10 light years away. (A light year is the distance that light travels in one year.) No matter how fast the astronaut travels, he cannot get there faster than light. The astronaut's trip must last more than 10 years as measured by an observer on Earth. An alien observer at the distant star is at rest relative to Earth. Therefore, the alien and the earthling observers have equivalent reference frames of space-time. On his trip, the astro-

Figure 6.5 Space-time map. An object moving at the speed of light has the same speed in both reference frames.

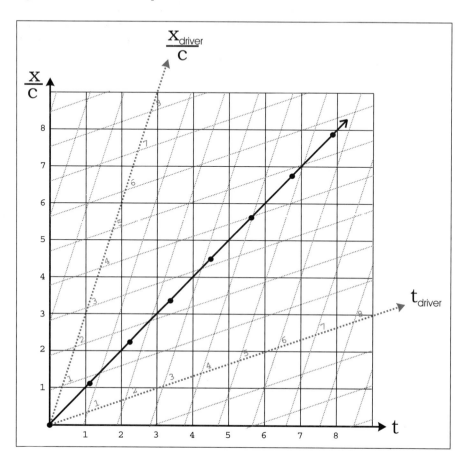

naut averages a speed that is 99% the speed of light. On the at-rest observer's watch, his trip lasts $t = (10 \text{ light years}) / 0.99 \times c \cong 10.1$ years. However, the astronaut's clock accumulates less time than the clocks at rest. To be exact, during the trip, he ages only $\tau = t/\gamma \cong 1.42$ years. He has traveled 10 light years, but aged less than 2 years! In contrast the earthling and the alien have both aged 10.1 years. From the astronaut's perspective he has traveled a velocity of 10 light years / 1.42 years $\cong 7c$, seven times the speed of light. This type of velocity is called *proper velocity*. Proper velocity is defined as proper distance divided by proper time. It can also be calculated by the formula $v_{proper} = \gamma \times v$.

Keep in mind, even though his proper velocity was greater than the speed of light, he could never win a race with a light beam. Light makes the trip in exactly 10 years earth time. If you could attach a clock to a photon, it would show that exactly zero seconds of proper time expired during the trip. Time freezes at the speed of light. For that reason, photons can be considered ageless. If you could take a trip at the speed of light, you would arrive at your destination instantly, as if you teleported. As your velocity approaches that of light, your proper velocity approaches infinity. You now understand another reason why you can't travel faster than the speed of light. Your proper velocity would be greater than infinity. To achieve this, time would need to run backwards for you.

Twin Paradox

A famous riddle of relativity is that of the "twin paradox." Assume that the astronaut mentioned earlier is your twin brother. He makes the trip to the distant star and then immediately turns back. Upon his arrival you are 20 years older but he is only about 3 years older. If motion is relative, why are you now of different ages? Doesn't this point out a contradiction in relativity? The answer is no. Uniform, straight-line motion is relative, but not all motion is relative. Upon starting his trip, your twin must have accelerated. He also must have experienced acceleration when he turned around at the distant star and when he slowed down on arriving back at earth. During these periods of acceleration, a force must have acted on the rocket and him. It is during these periods of acceleration that his space-time reference frame changes relative to yours. When he is coasting in uniform motion, the motion is relative, but during acceleration, he feels his "stomach drop" while you feel nothing. Accelerated motion is special and is not inherently relative. As always, it is during acceleration that nature "balances her books."

Space-Time, Momentum-Energy, and Other Four-Vectors

You already learned that space and time form a single four-dimensional structure. Events in space-time can be described by a four-dimensional quantity called a 4-vector: $\vec{x}_4 = [x,y,z,ct]$. The length of a 4-vector is calculated using the Minkowski metric, which you learned about earlier in the chapter. Part of the beauty of relativity is that all the fundamental quantities of physics are condensed using 4-vectors. For example, the proper time derivative of 4-position gives proper 4-velocity:

$$\vec{u}_4 = \frac{\partial \vec{x}_4}{\partial \tau} = [u_x, u_y, u_z, \gamma c]$$

where γ is the stretch factor. Multiplying 4-velocity by proper mass, m_o, produces 4-momentum:

$$\vec{p}_4 = m_o \vec{u}_4 = [p_x, p_y, p_z, \gamma m_o c]$$

Proper mass is also known as rest mass. Relativistic mass is defined as $m = \gamma m_o$. When you travel at high speed, your relativistic mass increases, but your *proper mass*, the mass that you feel in your reference frame, always stays the same. Substituting relativistic mass and multiplying by c, the fourth component of the four momentum becomes: mc^2, Einstein's famous equation for energy. Therefore, 4-momentum is often called the momentum-energy vector. The derivative of 4-momentum produces the 4-force, whose 4th component, when multiplied by c, gives the relativistic power. The laws of physics are dramatically condensed with relativity.

ELECTRIC FIELD AND MAGNETIC FIELD BECOME THE ELECTROMAGNETIC FIELD

The 4-vector concept also allows us to condense the mathematics of electromagnetics, and reveals the true nature of the electromagnetic field. By combining the magnetic vector potential with voltage into a four vector: $\vec{A}_4 = [A_x, A_y, A_z, V/c]$, a single electromagnetic potential is formed. Using this 4-potential, the electric and magnetic fields combine to form a single electromagnetic field, which can be represented by a mathematical object called a tensor. In the mathematical system called Clifford geometry, the electromagnetic field can be written simply as $F = E + icB$, where $i = \sqrt{-1}$. Within this mathematical system, Maxwell's equations condense to a single equation! (Refer to Baylis [1999].)

Even a Stationary Charge Has a Magnetic Field

A stationary charge has a magnetic field. This statement may seem to be a contradiction to what you have already learned about magnetic fields. Only moving charges produce magnetic fields, right? Let me explain.

Imagine yourself sitting in a wagon rolling along at a steady, constant speed. You have a charge in front of you. You measure it and it has an electric field but no magnetic field. You say I am wrong. But then you jump off the wagon, and you are on the ground. The charge is now moving relative to you. You measure again and you find that it has both an electric and a magnetic field. Let's say your friend Joe was standing by the side the whole time and your friend Bill stays in the wagon the whole time. For them nothing about the charge or its fields changed. Joe always observes the magnetic field, and Bill never observes it. When you were in the wagon you didn't observe the magnetic field, and then later you did. But nothing ever changed about the charge. The act of you jumping off the wagon didn't cause the charge to create a magnetic field. The act of you jumping off did not change anything about the charge, as witnessed by your two friends. The only way to reconcile this dilemma is to admit that the charge always had the magnetic field. You just couldn't observe it from your original perspective. The electric and magnetic fields are really different aspects of a single electromagnetic field. Here's an analogy. A three-dimensional cylinder looks like a two-dimensional circle from the end perspective and it looks like a two-dimensional rectangle from the side perspective. Only from other perspectives can you see the true three-dimensional nature. Similarly, when you are observing a charge from a perspective that is at rest with respect to the charge, the magnetic field is hidden.

Just as space and time are equal partners in the web of 4D space-time, the electric and magnetic field are equal partners in the 4D electromagnetic field. The voltage is analogous to time (1D), and the magnetic vector potential to space (3D). Now you understand why Maxwell's equations are correct in terms of special relativity. This fable of the charge and the wagon also explains why charges moving at constant velocity do not radiate. If they did, you could cause the charge to radiate by jumping off the wagon, but then your two friends would make contradicting observations. Relativity works behind the scenes of many electromagnetic phenomena.

THE LIMITS OF MAXWELL'S EQUATIONS

Maxwell's equations may have survived unscathed through the revolution of relativity, but the theory did not survive the discovery of quantum physics. The developments of quantum physics have shown that Maxwell's equations are an approximation, albeit a very good one. For many applications Maxwell's equations suffice.

In addition to not being able to correctly predict the observations of quantum phenomena such as thermal radiation and the operation of the laser, there is one inherent inconsistency in Maxwell's equations. This inconsistency involves the self-force of a charge and still has no known solution.

When a charge is accelerated, it radiates energy. The radiated energy comes from the force that causes the charge to move. For instance, imagine you have a charged ball and you move it back and forth. You must provide extra energy to move the field, which is observed as radiation. Because energy is required to move the field, the field has inertia or mass associated with it. The field mass is in addition to the mass of the particle itself (i.e., the mass that the particle would have were it not charged). Furthermore, since you must provide a force to move the field, the field must produce an equal but opposite force on you, as required by Newton's third law. How does the field convey this force? The field is a distributed entity. How can it produce an immediate and concentrated force on you, the source of the force? One way around this problem is to suppose that when the charge is accelerated, it encounters its own retarded or delayed field. The interaction of the charge with its own field causes a force to act back upon the charge and consequently a force acts upon you. This theory is in fact reasonable, but it causes other problems. Instead of talking about an arbitrary charge, consider an electron, the fundamental charge of nature. If the electron is assumed to be of finite, non-zero, dimensions, then under acceleration different parts of the electron would feel different amounts of force, and the electron would deform. The internal structure of the electron becomes unstable unless another, non-electromagnetic, force is invented to hold it together and to maintain its size and shape. To get around this problem, the electron can be assumed to be a point particle with no size or dimension. However, assuming that the electron is a point particle causes a slew of additional difficulties such as infinite charge density and the existence of simple problems that have either multiple solutions or no mathematical solution. The only known way out of this dilemma is through quantum electrodynamics.

QUANTUM PHYSICS AND THE BIRTH OF THE PHOTON

In 1900, Max Planck presented his theory on the quantization of thermal radiation energy levels. The prior "classical" theory of thermal radiation suffered from the "ultraviolet catastrophe." Simply stated, the ultraviolet castastrophe refers to the fact that the classical theory predicts that an object will radiate the majority of its thermal energy at ultraviolet frequencies and higher, contradicting the experimental fact that (save for objects of extreme temperature) thermal radiation peaks in the infrared. Moreover, the classical theory predicts that the radiated energy goes to infinity at high frequencies. To resolve the ultraviolet catastrophe of the classical theory of thermal radiation, Planck hypothesized that matter can only radiate heat energy in quantized packets of energy. Five years later, Einstein postulated that all electromagnetic radiation is quantized into particles now called photons. The energy of each radiating quanta or photon is proportional to its frequency: $E = hf$, where $h \cong 6.6 \times 10^{-34}$ joule-sec—Planck's constant. These two seemingly innocuous hypotheses opened the Pandora's box known as quantum physics. Science has never been the same since. Electromagnetic waves exhibit a so-called wave-particle duality of light. In some aspects, electromagnetic radiation acts as a distributed wave of energy. In other aspects, radiation acts as a localized particle.

Quantum Strangeness

"No one understands quantum physics." This is not the quote of a frustrated college student. Rather, it is the quote of Nobel Prize–winning physicist Richard Feynman. Feynman ironically won his Nobel Prize for his contributions to quantum theory. It was his work that led to a workable theory of *quantum electrodynamics (QED)*. He was not alone in his bewilderment. After the many strange consequences of quantum physics became apparent, several of its founders, including Planck, Einstein, and Schroedinger, not only declared that quantum physics was not understood, they outright rejected the theory. Nonetheless, quantum physics stands as one of the great achievements of science. Much of modern science, including the periodic table, physical chemistry, nuclear physics, lasers, superconductors, and semiconductor physics are based on quantum physics.

So why is quantum physics so strange? One example is the seemingly inherent randomness. At the quantum level, physics can only predict probabilities of possible outcomes. As an example, consider light impinging on a semireflective mirror, a mirror that is only partly silvered so that half of the light energy passes through and half of the light energy is reflected. From the particle point of view, half of the particles that make up the wave pass through and half are reflected. This statement seems reasonable at first. But as the light source is gradually reduced in amplitude, the number of photons is also reduced, until eventually only one photon at a time impinges upon the mirror. Yet even when a single photon hits the mirror, half of the time it passes through and half of the time it is reflected. Quantum physics explains this by saying that each photon has a random 50% probability of either reflection or transmission. There is no method to predict which way the photon will travel. The photon acts as if a coin toss determines which way it will travel.

Another example of quantum strangeness is that a particle can be in two places at the same time. The famous double slit experiment exemplifies the so-called non-local nature of quantum particles. Light of a single frequency is directed at two small, closely spaced slits in an otherwise opaque screen. Behind the screen is placed a photographic plate. After exposure, the plate shows a series of alternating light and dark bands. The bands are explained by the wavelike nature of the light. The light bands are regions of constructive interference and the dark bands are regions of destructive interference. The interference is caused by the fact that the two slits act as two sources of light waves that are slightly out of phase with each other. When the light amplitude is reduced to the single photon level, the photons map out the same pattern over time. What is amazing is that if one of the holes is covered, the pattern disappears. The pattern only appears when each photon travels through both holes. This is an example of how a particle, which is localized energy that exists in a single place, can also act as distributed wavelike energy that exists in more than one place at once. The unknown process by which a non-local wave turns into a localized particle is an example of what is called "collapse of the wavefunction," a topic that is still widely debated today.

Particles Are also Waves

In 1924 Louis de Broglie proposed, in his doctoral thesis, that matter also has a wave-particle duality. That is, all particles of matter also exhibit wavelike properties. The wavelength of the wave is related to the momentum of the particle, $\lambda = h/p$. Three years later, G.P. Thompson, whose father discovered the electron, showed experimentally that electrons behaved as waves. It is now accepted that all matter exhibits wavelike properties. The wavelike property of matter has been shown experimentally for atoms and molecules. For large everyday objects, such as baseballs, the wavelike property is negligible. For example, a 90 mph fastball has a wavelength on the order of 10^{-35} m—much too small to be noticeable.

In 1926 Erwin Schroedinger invented a system of mathematics to describe the physics of matter waves. His theory is based on several ad hoc postulates, which provide the basis for his wave mechanics. At first, scientists were unsure how to interpret the waves that are solutions to Schroedinger's equation. The Born interpretation, named after Max

Born, states that the square magnitude of the waves determines the probability of finding the particle at that location.

The Uncertainty Principle

The basis for quantum physics is that all electromagnetic energy is transferred in integer quantities of a fundamental unit, Planck's constant. Mathematically we can state this as $E = hf$, where E is energy, h is Planck's constant, and f is the frequency of the photon. To Einstein's dismay, it was this quantum theory that led directly to the Heisenberg Uncertainty Principle. In 1927, Werner Heisenberg published a paper stating quantized energy implies that all measurements have inherent uncertainty. Einstein expressed his dislike of this uncertainty as, "God does not throw dice," but was never able to disprove it. Specifically, the Heisenberg Uncertainty Principle states that we can never know both the exact position and momentum of any particle. Mathematically, the bounds on the error of our knowledge of position (Δx) and momentum (Δp_x) are related as

$$\Delta p_x \Delta x \geq \frac{h}{4\pi}$$

The uncertainty principle is not a limit set by the accuracy of our measuring equipment. It is a fundamental property of nature. This concept is straightforward to explain. To make a measurement of a particle, you must interact with it. Think about looking at small objects through a microscope. To be able to see what you're looking at, you shine a light source (or electron beam, etc.) on the object. While this may not have much consequence when measuring large objects like a baseball, such a simple task will drastically change the position and/or momentum of tiny objects like electrons. The only way to get around this problem would be if we could reduce the power of the light source to an infinitesimally small level. But Einstein's quantum theory puts a limit on how small the energy can be for a given frequency, f. The light wave must contain an integer multiple of energy, $E = Nhf$. To observe any object we must transfer at least $E = hf$ of energy, which will change the momentum of the object being observed. You may think that to avoid this problem, you could just decrease the frequency to a suitable level. In fact, the uncertainty in momentum can be reduced to an arbitrarily small value by reducing the frequency accordingly. However, the resolution of the light source is also dependent on the frequency. To increase the resolution, the frequency must be increased. In other words, you must increase the frequency of the light source to reduce the uncertainty in position of the object. The situation is a "catch-22". To decrease one type of uncertainty, you must increase the other uncertainty. Perfect certainty is impossible in quantum physics. We are left with bizarre consequences that even Einstein didn't foresee.

At about the same time that Schroedinger developed his wave mechanics, Heisenberg and several others developed matrix mechanics. This separate method for computing the results of quantum physics is actually mathematically equivalent to Schroedinger's wave mechanics. Today Schroedinger's wave mechanics and Heisenberg's matrix mechanics are known as quantum mechanics. A third mathematical formulation, known as the path integral method, was invented by Feynman in the 1940s.

THE QUANTUM VACUUM AND VIRTUAL PHOTONS

Now what about the energy in nonradiating electromagnetic fields, that is, the static field and the near field? Quantum physics states that any energy must consist of individual packets or quanta, but this implies that even the static field must consist of particles. In fact, the static field does consist of particles—virtual photons. To explain virtual photons, let's step further into the strange world of quantum physics.

If we think about the uncertainty principle from another point of view, it states that particles of small enough energy and short enough life-spans can exist but can never be measured. Mathematically, this is stated as

$$\Delta E \Delta t < \frac{h}{2\pi}$$

This allows for virtual particles to spontaneously appear and disappear as long as the uncertainty principle is obeyed. The law of conservation of energy can be violated during the fleeting existence of these virtual particles. These ephemeral particles can never be directly measured or observed, hence the term virtual particles. Any measurements or observations must obey the law of conservation of energy.

Now back to electromagnetic fields. The stored energy in an electromagnetic field allows for these virtual particles to be created and transport energy. It is these particles that carry the electromagnetic force in nonradiating fields. They have all the properties of the real photons that make up radiating fields with the exceptions being that they are fleeting in time and can never exist without their source being present.

An electron in free space is a good example. It is surrounded by a static electric field that stores energy. When there are no other charged particles present, the virtual photons that constitute the field will appear and disappear unnoticed, with no energy transferred. Now, if a second charge is placed near the electron, the virtual photons of the electron will transmit a force to this charge, and in a reciprocal manner the virtual photons from the field of the charged particle will transmit a force to the electron. This is the strange manner in which the electromagnetic force operates at the quantum level.

Quantum Physics, Special Relativity, and Antimatter

Roughly two years after Schroedinger invented wave mechanics, P.A.M. Dirac devised a combination of wave mechanics and special relativity. Dirac's equation predicted two additional strange consequences of quantum theory. First, any charged particle must exhibit angular momentum ("spin") and an associated magnetic field. Second, all charged particles have corresponding antiparticles. An antiparticle is identical to its corresponding particle except it has opposite charge. When a particle is combined with its antiparticle, they annihilate each other, producing photons with energy equivalent to $E = mc^2$. Soon after his prediction, the antiparticle of the electron, called a positron, was discovered. The positron has the same mass and size of an electron but is positively charged. Experiments also confirmed that particles and their antiparticle annihilate each other, producing energy in the form of photons. This experiment was important evidence for the validity of Einstein's relativistic energy equation and for the validity of quantum theory.

Matter Fields + Electromagnetic Fields = QED

An even stranger consequence of Dirac's equation was the prediction of matter fields. The existence of antimatter coupled with the uncertainty principle implies that particle/antiparticle pairs can appear out of the vacuum, similar to the virtual photons I described earlier in this chapter. Any charged particle such as an electron has a field whose energy is at all times surrounded by virtual electron/positron pairs, which arise from the quantum vacuum for a fleeting moment and then return to the vacuum. During their existence, however, they polarize. The virtual positron moves toward the real electron and the virtual electron moves away from the real electron. The end result is that the virtual electron/positron pairs shield the bare charge of the real electron, like the way charges polarize inside a conductor to reduce the E-field to zero. The virtual pairs tend to concentrate in the region close to the electron where the field is strongest and acts as an electrostatic shield. From a distance, the observed electron charge is much less than the bare charge. Moreover, any observations or interactions that occur very close to the real electron will be inside this cloud of virtual pairs, interacting with a larger value of charge. This cloud effect has been successful in predicting the Lamb shift of the inner electron orbitals of hydrogen. The electron is close enough to the hydrogen nucleus that it interacts with more of the bare charge from the nucleus than what is ordinarily considered as its observed value.

To summarize, the quantum theory of electromagnetics, known as quantum electrodynamics (QED), is based around the concepts of wave-particle duality and virtual particles in the quantum vacuum. The radiating field is a wave that consists of observable particles called photons. The nonradiating field consists of virtual photons which arise from the vacuum. The electron is considered to be a point particle with no dimension. Around the electron is a cloud of polarized electron/positron pairs. These virtual particle pairs have several effects. They shield the bare charge, causing the observed charge to be smaller than the bare charge of the electron. They also add mass to the electron because they follow the electron. Finally, they cause the electron's charge to appear as a cloud rather than a point charge. This cloud is condensed closely around the electron. As the cloud is penetrated, the observed charge increases. The bare charge of the point electron contains a negative infinite term. The electron itself also acts as a wave, in addition to being a particle.

QED was not and is not without its problems. Maxwell's equations predict an infinite mass if you assume that the electron is a point particle. This same problem occurs in QED. However, in the 1940s Feynman, Tomonaga, and Schwinger each solved the problem independently, for which they received the Nobel Prize in physics. To cancel the infinite term of the electromagnetic mass, the bare mass is defined to include a term that is equal but opposite in value to the electromagnetic mass. In other words, the bare mass contains a term that is infinite and negative. The positive infinity from the electromagnetic mass and the negative infinity from the bare mass cancel and the result is the finite mass that is observed in experiments. The technique, called *renormalization*, is not necessarily elegant but it works to incredible precision. QED is the most accurate physical theory that mankind has produced.

EXPLANATION OF THE MAGNETIC VECTOR POTENTIAL

In Chapter 3, I promised to give a conceptual meaning to the magnetic vector potential. This explanation involves the electrons of a superconducting circuit. In a superconductor, the electrons can move freely without colliding with the positive ions of the metal. Suppose that a DC current is applied to a superconducting wire. Without collisions, the electrons form a collective system that has a wave nature. In other words, the wave nature of the electrons dominates over the particle nature. Keep in mind that they form matter waves, not electromagnetic waves. The electromagnetic field and the current that result are still DC in nature. It is the location of the matter itself that forms a wave. In such a system the total momentum of an electron can be expressed as

$$\vec{p} = \frac{h}{2\pi}\vec{k}$$

Here k is the wave vector of the electron matter wave, whose magnitude is $2\pi/\lambda$, where λ is the wavelength of the matter wave. Hence, the total momentum is a quantum mechanical entity inversely proportional to the wavelength of the matter wave. Moreover, this total momentum is the sum of the electrodynamic momentum and the inertial momentum:

$$\vec{p} = q_o\vec{A} + m\vec{v}$$

Thus, the magnetic vector potential represents a true momentum in superconducting circuits. It is the electrodynamic momentum of the electron and its field. Mead (2000) provides excellent coverage of this topic.

THE FUTURE OF ELECTROMAGNETICS

Do we need something better than QED? The results of QED are based on a rather arbitrary process of subtracting infinities. The theory is wonderfully accurate but is not self-consistent. It also implies that the vacuum of empty space is actually a seething sea of infinite virtual particles. The foundation of QED is quantum mechanics, which is a system full of paradoxes that seem to defy understanding. We are left with a theory that works incredibly well, but cannot be understood conceptually. The traditional or Copenhagen interpretation of quantum mechanics states that we cannot understand what any of these quantum mechanical objects really are, we can just make observations and perform calculations. This interpretation was mainly developed and promoted by Niels Bohr. Most of the controversy of quantum mechanics revolves around how the wavefunction collapses; that is, how can an object that is in a superposition of many states change into an observable object that is always measured to be in a single state. Furthermore, is the process by which this happens inherently random? The Copenhagen interpretation basically states that this is beyond human knowledge—the question is unanswerable and irrelevant. Many alternative interpretations have been proposed, including the interpretation that consciousness is required (proposed by John von Neumann and Eugene Wigner) and the interpretation that includes a world of infinitely many universes (proposed by Hugh Everett III), as well as several others. In recent years, the increasing ability to create a variety of systems that exhibit quantum behavior has rekindled this area of research.

RELATIVITY, QUANTUM PHYSICS, AND BEYOND

Relativity and quantum physics were the major developments of physics in the 20th century. Scientists have yet to completely marry the two theories. Specifically, no theory of quantum gravity exists. One of the major goals for physicists in the 21st century is to develop a physical theory that encompasses all the known forces of nature, a so-called unified theory or theory of everything.

Relativity has repugnant assumptions but leads to a more simple and symmetric physics when compared to classical theory. It is also free of strange paradoxes, and can be understood conceptually. It was mainly the product of a single man, Albert Einstein, presented as a self-consistent theory.

In contrast, quantum physics started with a simple premise (electromagnetic wave energy is quantized), but has lead to horrible paradoxes. A span of about 50 years and the work of many physicists were required to develop the quantum theory of electrodynamics. Quantum physics is a theory that has developed over time through a chain of historical theoretical and experimental events.

A common connection between the two theories is that they both developed from questions concerning electromagnetic radiation. Relativity developed from questions about the speed of light, and quantum physics developed from questions about thermal radiation. Perhaps the theory that completely reconciles the two will be just as groundbreaking as each theory was itself.

BIBLIOGRAPHY AND SUGGESTIONS FOR FURTHER READING

Adler, C. G., "Does Mass Really Depend on Velocity, Dad?," *American Journal of Physics*, vol. 55, no. 8, pp. 739–743.

Albert, D. Z., "Bohm's Alternative to Quantum Mechanics," *Scientific American*, May 1994.

Baggot, J., *The Meaning of Quantum Theory*, Oxford: Oxford University Press, 1992.

Baylis, W. E., *Electrodynamics, A Modern Geometric Approach*, Boston: Birkhaeuser, 1999.

Blatt, F. J., *Principles of Physics*, 3rd Edition, Boston: Allyn and Bacon, 1989.

Boughn, S. P., "The Case of the Identically Accelerated Twins," *American Journal of Physics*, Vol. 57, No. 9, September 1989.

Callahan J. J., *The Geometry of Space Time: An Introduction to Special and General Relativity*, New York: Springer-Verlag, 2000.

Carrigan, R. A., Jr., and W. P. Trower, *Particle Physics in the Cosmos—Readings from Scientific American*, New York: W.H. Freeman and Company, 1989.

Carrigan, R. A., Jr., and W. P. Trower, *Particles and Forces at the Heart of the Matter, Readings from Scientific American*, New York: W.H. Freeman and Company, 1990.

Chiao, R. Y., P. G. Kwiat, and A. M. Steinberg, "Faster than Light?" *Scientific American*, August 1993.

Cramer, J. G., "The Transactional Interpretation of Quantum Mechanics," *Reviews of Modern Physics*, Vol. 58, p. 647, 1986.

Cramer, J. G., "An Overview of the Transactional Interpretation of Quantum Mechanics," *International Journal of Theoretical Physics*, Vol. 27, p. 227, 1988.

Davies, P. C. W., and J. R. Brown, *The Ghost in the Atom: A Discussion of the Mysteries of Quantum Physics*, Cambridge University Press, 1986.

Deutsch, D., and M. Lockwood, "The Quantum Physics of Time Travel," *Scientific American*, March 1994.

Duff, M. J., "The Theory Formerly Known as Strings," *Scientific American*, February 1998.

Eisberg, R., and R. Resnick, *Quantum Physics of Atoms, Molecules, Solids, Nuclei, and Particles*, 2nd Edition, New York: John Wiley & Sons, 1985.

Encyclopedia Britannica Inc., "Quantum Electrodynamics"; "Quantum Mechanics"; "Relativity," *Encyclopedia Britannica*, Chicago: Encyclopedia Britannica Inc., 1999.

Englert, B-G., M. Scully, and H. Walther, "The Duality in Matter and Light," *Scientific American*, December 1994.

Epstein, L. C., *Relativity Visualized*, San Francisco: Insight Press, 1988.

Feynman, R. P., *QED: The Strange Theory of Light and Matter*, Princeton, NJ: Princeton University Press, 1985.

Feynman R. P., *Quantum Electrodynamics*, New York: HarperCollins Publishers, 1998.

Feynman, R. P., Leighton, R. B., and Sands, M., *The Feynman Lectures on Physics*, Vol I, Reading, Mass.: Addison-Wesley, 1963.

Foster, J., and J. D. Nightingale, *A Short Course in General Relativity*, New York: Longman, 1979.

Georgi, H., "A Unified Theory of Elementary Particles and Forces," *Scientific American*, April 1981.

Greenwood, Stautberg M., "Use of Doppler-shifted Light Beams to Measure Time During Acceleration," *American Journal of Physics*, Vol. 44, No. 3, March 1976.

Greiner, W., *Relativistic Quantum Mechanics: Wave Equations*, 3rd Edition, Berlin: Springer-Verlag, 2000.

Greiner, W., *Quantum Mechanics an Introduction*, 4th Edition, Berlin: Springer-Verlag, 2001.

Greiner, W., and B. Muller, *Quantum Mechanics Symmetries*, Berlin: Springer-Verlag, 1989.

Greiner, W., J. Reinhardt (Contributor), and D. A. Bromley, *Quantum Electrodynamics*, 2nd Edition, Berlin: Springer-Verlag, 1996.

Gribbin, J., *In Search of Schrodinger's Cat: Quantum Physics and Reality*, New York: Bantam, 1984.

Gribbin, J., *Schrödinger's Kittens and the Search for Reality: Solving the Quantum Mysteries*, Boston: Little, Brown & Co., 1995.

Griffiths, D. J., *Introduction to Electrodynamics*, 3rd Edition, Upper Saddle River, NJ: Prentice Hall, 1999.

Guilini, D., E. Joos, C. Kiefer, J. Kupsch, I. O. Stamatescu, and H. D. Zeh, *Decoherence and the Appearance of a Classical World in Quantum Theory*, Berlin: Springer-Verlag, 1996.

Hawking, S. W., and R. Penrose, "The Nature of Space and Time," *Scientific American*, July 1996.

Horgan, J., "Quantum Philosophy," *Scientific American*, July 1992.

Kwiat, P., H. Weinfurter, and A. Zeilinger, "Quantum Seeing in the Dark," *Scientific American*, November 1996.

Liboff, R. L., *Introductory Quantum Mechanics*, 3rd Edition, North Reading, Mass.: Addison-Wesley Publishing Company, 1998.

Lounesto, P., *Clifford Algebras and Spinors, London Mathematical Society Lecture Note Series* 239, Cambridge, UK: Cambridge University Press, 1997.

Mead, C. A., *Collective Electrodynamics—Quantum Foundations of Electromagnetism*, Cambridge, Mass.: MIT Press, 2000.

Milonni, P. W., *The Quantum Vacuum—An Introduction to Quantum Electrodynamics*, San Diego: Academic Press, 1994.

Moore, T. A., *A Traveler's Guide to Space Time—An Introduction to the Special Theory of Relativity*, New York: McGraw-Hill, 1995.

Perrin, R., "Twin Paradox: A Complete Treatment from the Point of View of Each Twin," *American Journal of Physics*, Vol. 47, No. 4, April 1979.

Purcell, E. M., *Electricity and Magnetism*, 2nd Edition, McGraw-Hill, 1985.

Rae, A. I. M., *Quantum Physics: Illusion or Reality?*, Cambridge, UK: Cambridge University Press, 1998.

Rothman, M. A., "Things That Go Faster than Light," *Scientific American*, July 1960.
Schutz, B. F., *A First Course in General Relativity*, New York: Cambridge University Press, 1985.
Schwartz, M., *Principles of Electrodynamics*, New York: Dover Publications Inc, 1972.
Semon, M. D., and J. R. Taylor, " Thoughts on the Magnetic Vector Potential," *American Journal of Physics*, Vol. 64, No. 11, November 1996.
Shanker, R., *Principles of Quantum Mechanics*, 2nd Edition, New York: Plenum Press, 1994.
Shapiro, I. I., "A Century of Relativity," *Reviews of Modern Physics*, Vol. 71, No. 2, 1999.
Smolin, L., *Three Roads to Quantum Gravity*, New York: Basic Books, 2001.
't Hooft, G., "Gauge Theories of the Forces between Elementary Particles," *Scientific American*, June 1980.
Taylor, E. F., and J. A. Wheeler, *Spacetime Physics Introduction to Special Relativity*, 2nd Edition, New York: W.H. Freeman and Company, 1992.
Vanderlinde, J., *Classical Electromagnetic Theory*, New York: John Wiley & Sons, 1993.
Will, C., *Was Einstein Right?: Putting General Relativity to the Test*, New York: Basic Books, 1986.
Wu, T-Y, and W-Y P. Hwang, *Relativistic Quantum Mechanics and Quantum Fields*, Singapore: World Scientific, 1991.
Yam, P., "Bringing Schrodinger's Cat to Life," *Scientific American*, June 1997.
Zeh, H. D., "On the Interpretation of Measurement in Quantum Theory," Foundations of Physics Vol. 1, p. 69–76, 1970.
Zeilinger, A., "Quantum Teleportation," *Scientific American*, April 2000.
Zurek, W., "Decoherence and the Transition from Quantum to Classical," *Physics Today*, October 1991.

Web Resources

http://www.decoherence.de/
http://setis.library.usyd.edu.au/stanford/entries/qm-everett/
http://www.umsl.edu/~fraundor/a1toc.html
http://math.ucr.edu/home/baez/physics/faq.html
http://web.mit.edu/afs/athena.mit.edu/user/r/e/redingtn/www/netadv/welcome.html
http://www.treasure-troves.com/physics/
http://www.yourphysicslink.com/quantummechanics/quantumfieldtheory/
http://www.superstringtheory.com/

7 THE HIDDEN SCHEMATIC

Typically, schematic diagrams show resistors, capacitors, inductors, and wires as ideal components. In many cases, components can be considered as ideal. However, at high frequencies such approximations are often no longer valid. The frequency-dependent departures from ideality are mainly due to parasitic capacitances and parasitic inductances. Electromagnetic theory dictates that any two conductors will have a capacitance between them and that any conductor used for carrying current will have inductance. Parasitic capacitance and inductance create reactive impedance that varies with frequency. For a capacitor, the impedance is $Z_c = -j/2\pi fC$. At DC, capacitive impedance is infinite—an open circuit. Capacitive impedance decreases with frequency. For an inductor, impedance is $Z_L = j2\pi fL$. At DC, inductive impedance is zero—a short circuit. Inductive impedance increases with frequency. These parasitic impedances cause real components to behave differently at high frequencies. The parasitic impedances of components are present in every real component but are typically not shown in their schematic symbols, hence the phrase *hidden schematic*.

In addition to the frequency-dependent parasitic elements, there exist other non-ideal characteristics of components. All conductors, including wires and component leads, have a non-zero resistance. At times this resistance must be accounted for in the design process. The characteristics of all components also vary with temperature. Components made from different materials will often display varying degrees of sensitivity to temperature. Temperature dependence must be accounted for in any real-world design. Finally, individual components will vary from the nominal value. The amount by which a component can be expected to vary from the nominal value is called *tolerance*.

THE NON-IDEAL RESISTOR

There are three common types of resistors: carbon resistors, wire-wound resistors, and metal film resistors. Carbon resistors are inexpensive

Figure 7.1 The hidden schematic of a real-world resistor includes lead inductance (L_{lead}) and parasitic capacitance (C_p). Most often (except in the case of very low resistance values) it is the parasitic capacitance that limits the high-frequency performance.

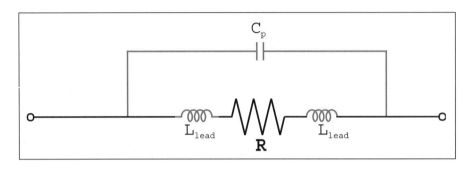

components that can use the low conductivity of carbon to create resistance. Wire-wound resistors are simply a very long wire, wound into a tight form. Metal film resistors are thin films that create resistance due to their small cross-sectional area. Wire-wound and metal film resistors are available in tighter tolerances and lower temperature coefficients than are carbon resistors. Low price is typically the only benefit of carbon resistors. Another consideration for resistors is their power-handling capability. If too much current is forced through a resistor, it will become too hot and will burn up or experience other permanent damage. Resistors are therefore given power ratings.

Real resistors also have frequency limitations. An equivalent circuit is shown in Figure 7.1. The leads of a resistor create a capacitance that is in parallel to the resistance. The leads of the resistor, and the resistive material itself create a series inductance. Both parasitic effects limit the frequency range of the resistor. For large resistances, the capacitance dominates the high-frequency response, shunting out the resistance and reducing the effective impedance of the resistor. Figure 7.2 shows the frequency response of two resistors.

The resistance remains constant until a corner frequency, defined by $f = 1/(RC)$. The frequency performance of a resistor, therefore, is worse for large resistances and for large parasitic capacitances. For this reason, high-resistance resistors are not used at high frequencies. To improve performance, the leads of a resistor can be shortened, reducing the parasitic capacitance. Surface-mount chip resistors, which are rectangular

THE NON-IDEAL RESISTOR

Figure 7.2 Simulated frequency-dependent behavior of typical axial (through-hole mount) resistors: A) $R = 50\,\text{kohm}$, $L_{lead} = 8\,\text{nH}$, $C_p = 0.3\,\text{pF}$; B) $R = 5\,\text{ohm}$, $L_{lead} = 8\,\text{nH}$, $C_p = 0.3\,\text{pF}$.

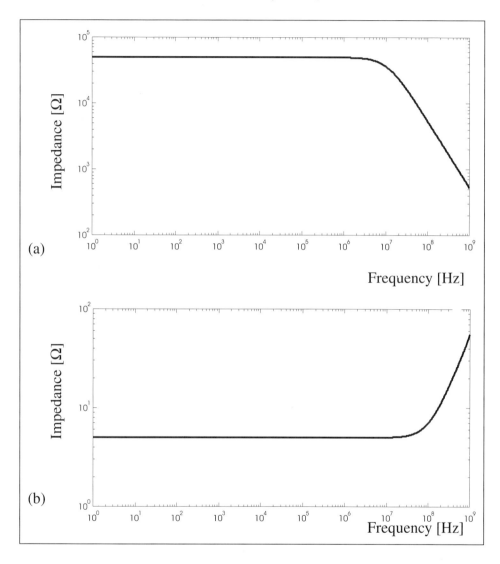

blocks of resistive material with metal ends, reduce capacitance greatly because they do not have extended leads.

For small-valued resistors, the parasitic inductance dominates the high-frequency response, increasing the effective impedance. Wire-wound resistors have notoriously high inductance, and cannot be used at high frequencies. Therefore, at high frequencies the very high- and very low-valued resistors will change the most. It turns out that 50 ohms, the most common characteristic impedance for use at high frequencies, serves as a good gauge for determining whether or not a resistor is very high or very low. A general, but somewhat arbitrary, guideline to use for selecting chip resistors is

$$\frac{f}{3 \times 10^{10}\,\text{Hz}} \leq \frac{R}{50\,\Omega} \leq \frac{3 \times 10^{10}\,\text{Hz}}{f}$$

where f is the frequency of operation in Hz and R is resistance in ohms. Of course this is a rule of thumb and is not a substitute for measurements or simulation.

THE NON-IDEAL CAPACITOR

An ideal capacitor has a reactive impedance of $Z_c = -j/(2\pi fC)$. All real capacitors have several parasitic elements, as shown in Figure 7.3.

The leads of the capacitor have inductance and resistance. In addition, the dielectric material between the plates does not have infinite

Figure 7.3 The hidden schematic of a real-world capacitor includes lead inductance (L_{lead}), lead resistance (R_{lead}), dielectric leakage (R_{DC}), and dielectric "frictional" loss (R_{AC}). The resistances are often combined and called Equivalent Series Resistance (ESR). Dielectric losses (and hence ESR) are typically frequency dependent.

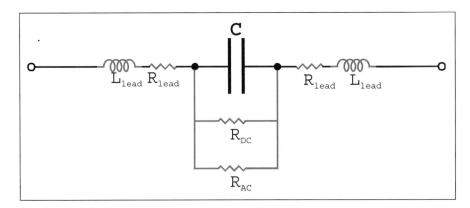

resistance, although the actual DC or leakage resistance is typically quite large. In addition to the DC resistance there is an AC loss, which can be represented as a resistance. The AC loss corresponds to microscopic frictional loss, which arises from polarization of charge as the voltage changes. This loss translates to heat and is very important to consider when using capacitors for high-current applications such as power supplies. If too much AC current passes through the capacitor, it will heat up, limiting its useful lifetime. Manufacturers often specify this resistance as ESR (Equivalent Series Resistance), which lumps all the resistances together. Ceramic capacitors typically have low ESRs, but they require too large a volume to be practical for large capacitance values. Aluminum electrolytic and tantalum electrolytic capacitors are available in large capacitance values but have higher ESRs. Of the two types, tantalum electrolytic capacitors have lower ESRs. Figure 7.4 shows the frequency response of several capacitors.

For high-frequency design, the parasitic series inductance poses the largest problem. The inductance creates a resonant frequency at $f_o = 1/\sqrt{(LC)}$. Above this frequency, the impedance of the capacitor increases with frequency; in other words, the component acts like an inductor above the resonant point. There are two options to increase the resonant frequency: 1) reduce the parasitic inductance or 2) use a smaller value of capacitance. From the second statement, you can assume that large values of capacitance are not useful at high frequencies.

Different dielectric materials also have different temperature dependences. Ceramic materials have less temperature dependence than do electrolytic capacitors. The last parameter to consider in choosing capacitors is the working voltage. Above the working voltage the dielectric will be damaged. At high enough voltages the dielectric will break down and short circuits will form between the plates.

THE NON-IDEAL INDUCTOR

In general, inductors are more problematic than capacitors. The circuit model for a real inductor is shown in Figure 7.5. The parasitic elements are: 1) resistance within the leads and the wire of the inductor, 2) the capacitance between the leads and between the loops of wire, and 3) the equivalent resistance corresponding to core losses (if the inductor uses a ferromagnetic core). The parasitic capacitance forms a resonant circuit with the inductance, with a resonant frequency at $f_o = 1/\sqrt{(LC)}$. Above this frequency, the impedance of the inductor decreases with frequency; in other words, the component acts like a

Figure 7.4 A) Frequency response of a 0.1 µF surface-mount, size 0805 (0.08" × 0.05") capacitor (L_{lead} = 0.73 nH). B) Frequency response of a 0.015 µF surface-mount, size 0805 (0.08" × 0.05") capacitor (L_{lead} = 0.88 nH). Above the self-resonant frequency (SRF) of ~40 MHz, the device acts like a 0.88 nH inductor. C) Frequency response of a 0.001 µF surface-mount, size 0805 (0.08" × 0.05") capacitor (L_{lead} = 0.77 nH). Above the self-resonant frequency (SRF) of ~200 MHz, the device acts like a 0.77 nH inductor. Plots were created with muRata's MCSIL software (http://www.murata.com/).

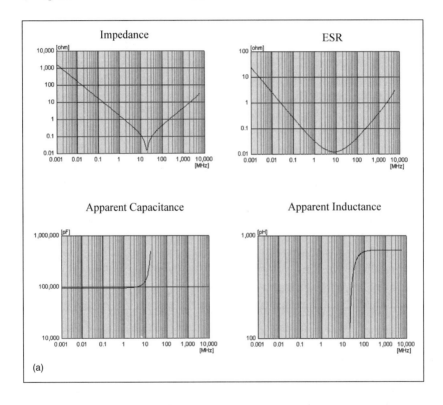

capacitor above the resonant point. Again, there are two options to increase the resonant frequency: 1) reduce the parasitic capacitance or 2) use a smaller value of inductance. Large values of inductance, thus, are not practical at high frequencies.

If the inductor contains a ferromagnetic core, the core losses will also limit frequency response. The core losses arise from hysteresis losses and from eddy currents within the core. The situation is compli-

Figure 7.4 *Continued.*

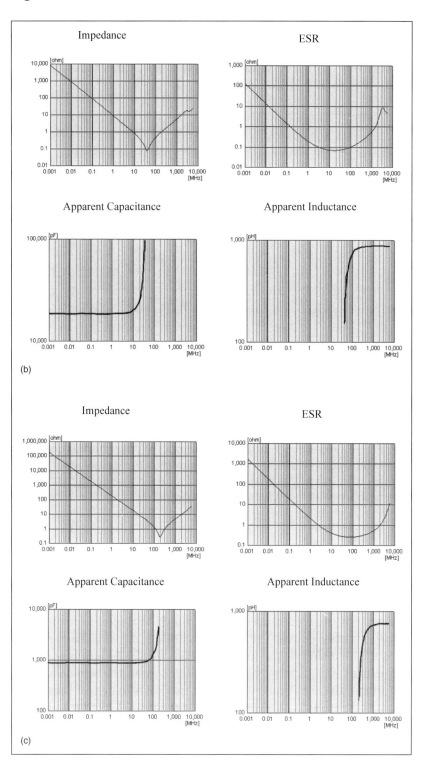

Figure 7.5 Hidden schematic of a real-world inductor includes lead resistance (R_{lead}), core loss (R_{core}), and parasitic capacitance (C_p) resulting from the leads and windings. Core losses are typically frequency dependent.

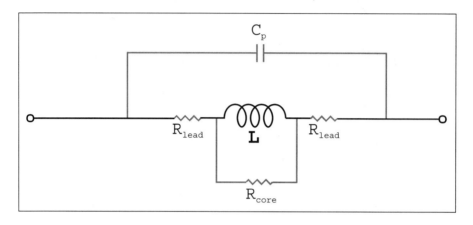

cated by the fact that hysteresis losses are nonlinear and that eddy current losses increase with frequency. Figure 7.6 plots the frequency response of two non-ideal inductors.

All real inductors are limited as to how much current they can carry. This limitation stems from the resistance of the conductors that make up the inductor and its leads. Inductors with ferromagnetic cores are further limited in that the current must be kept below the saturation level for the device to operate properly.

NON-IDEAL WIRES AND TRANSMISSION LINES

In circuit analysis, wires are usually considered to be ideal short circuits. All wires have resistance, which can be a problem at any frequency if the wire gauge (or trace width in the case of printed circuit traces) is not chosen properly. In addition, all wires have inductance. At low frequencies, the inductance can usually be ignored. At high frequencies, the inductance greatly affects how the wires carry signals. Furthermore, all circuits use at least two wires to deliver the signal from source to load. Typically one wire is designated for the source current and the other is designated for the return current and is often grounded.

Figure 7.6 A) Frequency response of a ferrite-core, 100 nH surface-mount inductor (C_p = 1 pF; R_{DC} = 0.26 ohm). Above about 80 MHz, the ESR increases dramatically due to ferrite losses in the core. The self-resonant frequency occurs at about 500 MHz. Above the self-resonant frequency, the device acts like a ~1 pF capacitor.

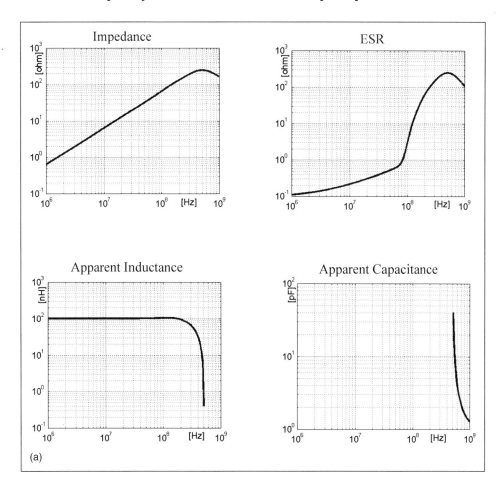

(a)

Like any two conductors, these two wires have a capacitance between them. A simple circuit is shown in Figure 7.7, showing the parasitic elements of the wires.

To make the problem of parasitic inductance and capacitance manageable at high frequencies, transmission lines are used. A transmission

Figure 7.6 B) Frequency response of an air-core, 100 nH surface-mount inductor (C_p = 0.25 pF; R_{DC} = 0.38 ohm). The self-resonant frequency occurs at about 1 GHz. Above the self-resonant frequency, the device approaches the behavior of a ~0.25 pF capacitor.

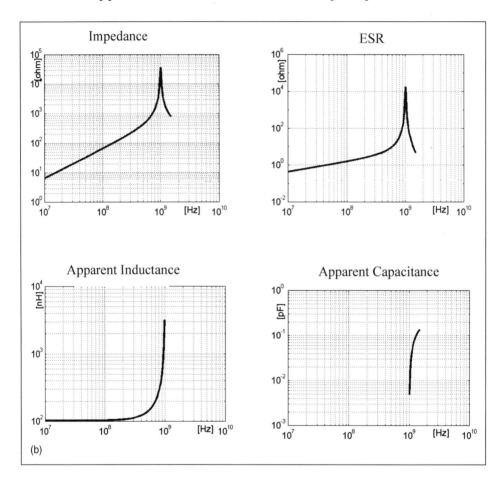

line is a set of conductors with uniform spacing. Some examples are coaxial cable (used for cable TV and lab instruments), twisted pair cable (used for telephone lines and computer network cables), ribbon cables (used to connect circuit boards inside computers), twin lead cables (used for TV antennas), microstrip and stripline (used on printed circuit boards). Transmission lines serve to guide the electromagnetic signals predictably and to reduce radiation.

Figure 7.7 A) A simple circuit consisting of a voltage source connected to a load resistor. B) The hidden schematic for the wires includes resistance (R_{wire}), inductance (L_{wire}), and parasitic capacitance (C_p).

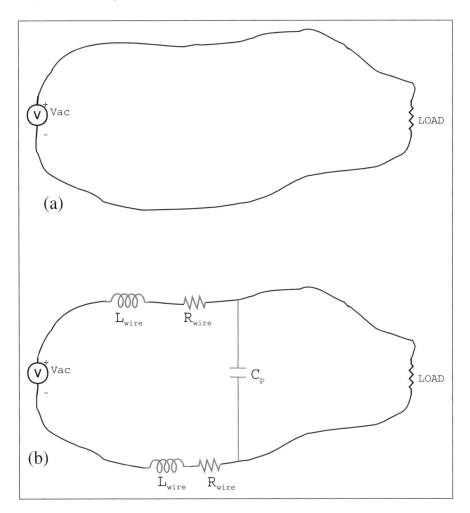

OTHER COMPONENTS

Resistors, capacitors, and inductors are not the only components that change behavior at high frequency. All components have parasitic behavior. Transformers are basically coupled inductors, and are the most complicated of the passive components. Semiconductor parts such as diodes, transistors, and amplifiers also have parasitic capacitance and inductance. In addition, the electronic behavior of the semiconductor material itself plays a role in high-speed applications.

MAKING HIGH-FREQUENCY MEASUREMENTS OF COMPONENTS

Accurately measuring the high-frequency response of components is not an easy task. It requires the use of a vector network analyzer (VNA) and a test fixture (a device in which to place the components during testing). The difficulty arises in isolating the component response from the cables, connectors, and test fixture used to connect the component to the network analyzer. Very accurate calibration is therefore needed. If the standard SOLT (Short Open Load Thru) calibration technique is used, then an accurate calibration kit (set of short, open, load, and thru standards) must be constructed for the test fixture. Furthermore, the calibration kit parameters for your test fixture must be entered manually into the network analyzer before the actual calibration of the instrument is performed. Determining the calibration kit parameters is a task that involves making an accurate high-frequency model of your test fixture and connectors. Another option is to use the TRL (Transmission Reflection Load) calibration technique. Calibration kit parameters are not needed for this technique. The most accurate and easiest route to take is to purchase component test fixtures from a company such as Agilent Technologies (formerly the test and measurement division of Hewlett-Packard). To make measurements on an IC wafer, microwave probe stations such as those manufactured by Cascade Microtech can be used.

RF COUPLING AND RF CHOKES

RF amplifiers typically require that you AC couple to the input and output. In other words, you must use a capacitor in series with the input and output, otherwise you will disrupt the DC bias point of the amplifier. For these applications, I typically select the capacitor such that the magnitude of its impedance is about 1 ohm at the signal frequency, $C = 1/2\pi f$. Setting the capacitor to this value balances the two design considerations involved. First, the capacitor should be large enough that it presents a very low-ohm connection for the signal to travel through. Second, the capacitor should be small enough that its self-resonant frequency is above the frequency range of the signal. The value of 1 ohm is a good compromise for these two requirements.

RF amplifiers also often require that you place an inductor in series with the power input to prevent the signal from leaking onto the power supply. Such an application of an inductor is often called an RF choke. Ideally, you would like the impedance of the RF choke to be as large as possible at the frequency of interest. You also need to be sure that the signal frequency is below the self-resonant frequency of the inductor. Finally, you must be sure the inductor can handle the DC current without saturating or being damaged. Typically, you should aim for an impedance of 500 ohms or greater at the signal frequency to achieve good results.

COMPONENT SELECTION GUIDE

Parasitic capacitance and inductance limit the frequency response of all components including wires. The frequency response of the materials themselves can also play a role; examples are dielectric materials in

Table 7.1 Selection Considerations for Real-World Components

Component	Considerations
Resistors	Tolerance
	Power rating
	Temperature coefficient
	Parallel capacitance
	Series inductance
Capacitors	Tolerance
	Voltage rating
	Temperature coefficient
	Series inductance (or resonant frequency)
	Equivalent series resistance (ESR)
Inductors	Tolerance
	(Thermal) Current rating
	Parasitic capacitance (or resonant frequency)
	Series resistance
	Saturation current rating (if ferromagnetic core)
	Cores loss frequency response (if ferromagnetic core)

capacitors, core materials in inductors, and semiconductor materials in active elements. In general, components with smaller physical size will have smaller parasitics and therefore better high-frequency performance. Keeping resistance, capacitance, and inductance values small also helps at high frequency. SMD (Surface Mount Device) components allow for lumped element design on printed circuit boards (PCBs) up to several GHz. Integrated circuits, which allow very small components to be created, are needed at higher frequencies. Another option for high-frequency electronics is the use of distributed elements as described in Chapter 10. Table 7.1 gives some general selection rules for common passive components.

BIBLIOGRAPHY

Agilent Technologies, *Understanding the Fundamental Principles of Vector Network Analysis*, Agilent Technologies, AN 1287-1.

Agilent Technologies, *Accessories Selection Guide for Impedance Measurements*, Agilent Technologies, April 2000.

Bowick, C., *RF Circuit Design*, Carmel, Indiana: SAMS Publishing, 1995.

Breed, G. A., "Inductor Behavior at Radio Frequencies," *RF Design*, February 1996.

Hamilton, N., "A Cheaper Alternative to Measuring Complex Impedance," *RF Design*, January 1999.

Paul, C. R., *Introduction to Electromagnetic Compatibility*, New York: John Wiley & Sons, 1992.

Pease, R. A., *Troubleshooting Analog Circuits*, Newton, Mass.: Butterworth–Heinemann, 1991.

Ramo, S., J. R. Whinnery, T. Van Duzer, *Fields and Waves in Communication Electronics*, 2nd Edition, New York: John Wiley, 1989.

Rhea, R. W., "A Multimode High-Frequency Inductor Model," *Applied Microwave & Wireless*, November 1997.

Shrader, R. L., *Electronic Communication*, 5th Edition, New York: McGraw-Hill, 1985.

Williams, T., *The Circuit Designer's Companion*, Newton, Mass.: Butterworth–Heinemann, 1991.

8 TRANSMISSION LINES

As wireless designs become more prevalent and as digital designs reach higher and higher frequencies, a thorough understanding of transmission line theory is becoming increasingly important. With the aid of graphical representations of analog and digital signals, you can gain a solid intuitive understanding of transmission lines. Moreover, this approach requires little mathematics. Unfortunately, many engineers leave school having been exposed to transmission lines only during a few lectures in an electromagnetic fields class. In such classes, transmission line theory is taught with wave equations and a lot of difficult calculus. You have probably heard that transmission line effects become apparent at higher frequencies, but rarely does anyone explain why. Why are transmission line effects usually noticeable only at high frequencies? What happens at low frequencies? What are the definitions of "high" and "low"? In practice, you can more easily and completely grasp transmission line theory just by understanding the basic physics.

THE CIRCUIT MODEL

As the name implies, a transmission line is a set of conductors used for transmitting electrical signals. In general, every connection in an electric circuit is a transmission line. However, implicit in most discussions of transmission line theory is the assumption that the lines are uniform. A uniform transmission line is one with uniform geometry and materials. That is, the conductor shape, size, and spacing are constant, and the electrical characteristics of the conductors and the material between them are uniform. Some examples of uniform transmission lines are coaxial cables, twisted-wire pairs, and parallel-wire pairs. For printed circuit boards (PCBs), the common transmission lines are strip-line and microstrip.

In a simple transmission line circuit, a source provides a signal that is intended to reach a load. Figure 8.1A shows the transmission line as

Figure 8.1 Three models for a transmission line: A) an ideal pair of wires; B) a lumped circuit model that includes the non-ideal characteristics of the wires; C) a simplified, lossless, lumped circuit model.

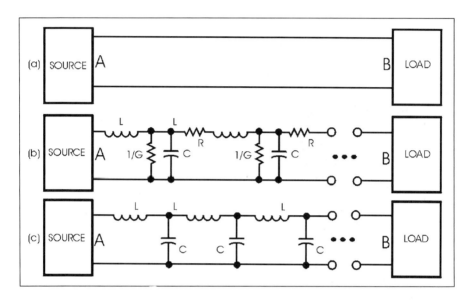

a pair of parallel conductors. In basic circuit theory, you assume that the wires making up the transmission line are ideal and hence that the voltage at all points on the wires is exactly the same. In reality, this situation is never quite true. Any real wire has series resistance (R) and inductance (L). Additionally, a capacitance (C) exists between any pair of real wires. Moreover, because all dielectrics exhibit some leakage, a small conductance (G) (that is, a high shunt resistance) exists between the two wires. You can model the transmission line using a basic circuit that consists of an infinite series of infinitesimal R, L, and C elements (Figure 8.1B). Because the elements are infinitesimal, the model parameters (L, C, R, and G) are usually specified in units per meter. To simplify the discussion, you can ignore the resistances. Figure 8.1C shows the resulting LC circuit. A transmission line that is assumed to have no resistance is a *lossless* transmission line.

Notice several important points. First, with the LC model, points A and B may be at different potentials. Second, a signal transmitted from the source charges and discharges the line's inductance and capacitance. Hence, the signal does not arrive instantly at Point B but is delayed. Last, the impedance at points A and B and each node in between depends not just on the source and load resistance, but also on the LC values of the transmission line. How does this circuit react at different frequencies? Recall that inductive and capacitive reactance depend on frequency. At low frequencies, the LC pairs introduce negligible delay and impedance, reducing the model to a simple pair of ideal wires. At higher frequencies, the LC effects dominate the behavior, and you cannot ignore them.

CHARACTERISTIC IMPEDANCE

The simple circuit of Figure 8.2 demonstrates the behavior of a transmission line. In this circuit, a 5 volt battery is connected to a resistor through a transmission line that is modeled with a series of four L-C sections. In reality, a real transmission line is an infinite series of infinitesimal inductors and capacitors. However, this simplified model serves as a good heuristic tool. Between the battery and the transmission line is a mechanical switch, which is initially open. In the initial state there is no voltage on the line or the load, and no current flows. Immediately after the switch is closed, current flows from the battery into the transmission line. At this point, the current does not reach the load. Instead the current is diverted by the first capacitor. The capacitor continues to sink charge until it reaches the 5 volts of the battery. During this process some energy is also transferred to the magnetic field of the inductor. As the voltage on the capacitor starts to climb, charge starts to trickle to the second stage.

During the charging of the capacitors and inductors, no current reaches the load. Therefore the impedance that the battery "sees" is solely dependent on the value of the inductors and capacitors. This impedance is referred to as the *characteristic impedance* of the transmission line, and is easily calculated using the formula,

$$Z_o = \sqrt{\frac{L}{C}}$$

As you can infer from the formula, a lossless transmission line has a purely real characteristic impedance (i.e., a resistance). For the curious reader, I include the equation for a lossy transmission line:

$$Z_o = \sqrt{\frac{R + i\omega L}{G + i\omega C}}$$

Figure 8.2 A conceptual demonstration of transmission line delay and characteristic impedance. A battery is connected to a resistor through a transmission line, which is represented by a series of L-C stages. A) The switch is open. No current flows and the voltage of the transmission line is zero everywhere. $I = 0$. B) The switch is closed. Current flows and starts charging the first L-C stage. $I = V_{battery}/Z_o$, where $Z_o = \sqrt{L/C}$. C) The battery has charged the first stage and is now charging the second stage. $I = V_{battery}/Z_o$. D) The battery has charged the entire transmission line, and now current flows through the load resistor. $I = V_{battery}/R_{load}$.

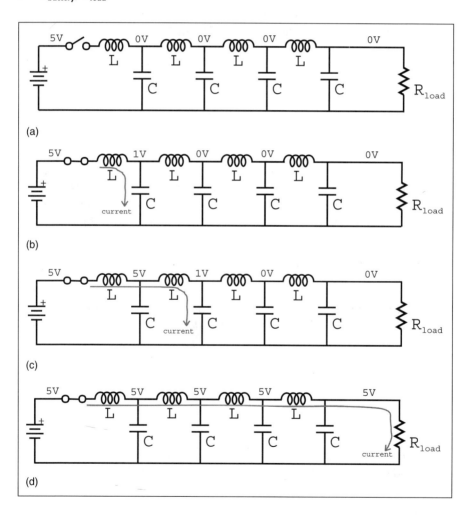

where $i = \sqrt{-1}$.

For the time that the transmission line is charging, energy is drawn from the battery and transferred to the magnetic field of the inductors and the electric field of the capacitors. Typically a transmission line charges much faster than is observable by human abilities. However, the time it takes to charge a transmission line is directly proportional to its length. If you built a very long transmission line, for example a coaxial cable from the earth to the moon, these effects would be observable on a time scale perceptible even to humans. Suppose the characteristic impedance of this extremely long cable is 50 ohms and the resistance of the load at the other end of the cable (the end at the moon) is 1000 ohms. Assuming that the coaxial cable is filled with a Teflon cable, it will take about two full seconds to charge the transmission line. Suppose you connect an analog ohmmeter to the line. For approximately four seconds the meter will read 50 ohms—the characteristic impedance of the cable. Only after the transmission line is fully charged and all resonant oscillations (often called ringing) die out will the meter steadily read 1000 ohms. Characteristic impedance is a very real effect, not just some textbook parameter.

In contrast to this extremely long cable, a cable of 1 meter in length will take only a few nanoseconds to charge. Such small time scales are negligible to human perception and are also inconsequential to slow signals such as audio. However, these time scales are not small in reference to RF systems and modern computers. For this reason, transmission line effects become noticeable and problematic at high frequencies.

THE WAVEGUIDE MODEL

An equivalent method of characterizing transmission lines describes a transmission line as a guide for electromagnetic waves. With this method, the source sends the electromagnetic signal, which consists of a coupled voltage wave and current wave, to the load (Figure 8.3). The voltage wave corresponds to the electric field and the current wave corresponds to the magnetic field of the wave. The transmission line, which effectively acts as a transmission medium, guides the signal along the way. The signal travels through this medium at the speed of light within that medium.

You can calculate the speed of light, v, in a transmission line from the permittivity (ε) and permeability (μ) of the dielectric between the conductors,

Figure 8.3 The waveguide model of the transmission line represents the signal as a voltage wave and a current wave that both travel along the line at velocity, *v*. This velocity is the speed of light of the dielectric between the wires. In the case of air, this velocity is approximately that of the speed of light in a vacuum. The voltage wave is equal to the current wave multiplied by the constant, Z_o, which is the characteristic impedance of the line.

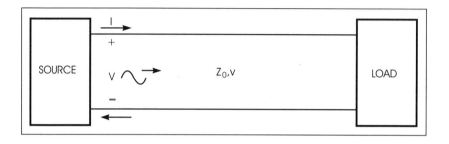

$$v = \sqrt{\frac{1}{\varepsilon\mu}}$$

For example, if Teflon separates the pair of wires that makes up the transmission line, the wave travels at the speed of light in Teflon, which is approximately 70% of the speed of light in a vacuum. ($v_{TEFLON} \cong 0.7c$, where *c* is the speed of light in a vacuum.) As the signal travels along the transmission line, the voltage wave defines the voltage at each point, and the current wave defines the current at each point.

Figure 8.4 shows a pulse signal traveling along a transmission line, with the voltage and current values at each point. Along the entire length of the line, the ratio of the voltage to the current is constant. This ratio is the characteristic impedance, Z_o, and is defined by the geometry of the line and the permittivity and permeability of the dielectric. The characteristic impedance equations are often complex. For example, for a coaxial cable transmission line,

$$Z_o \cong \frac{60}{\sqrt{\varepsilon_r}} \ln\left[\frac{D}{d}\right]$$

where *d* is the diameter of the inner conductor, *D* is the inside diameter of the outer conductor, and ε_r is the relative dielectric constant of the material.

Figure 8.4 This figure shows two snapshots in time of a signal pulse traveling along a transmission line. Both the voltage and current waves are shown.

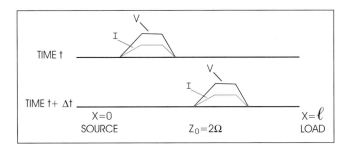

RELATIONSHIP BETWEEN THE MODELS

Now you have two models for a transmission line: a circuit comprising infinitesimal inductances and capacitances with parameters L and C, and a waveguide for signals with parameters v and Z_o. The following equations relate the parameters of the two lossless models:

$$Z_o = \sqrt{\frac{L}{C}}$$

and

$$v = \sqrt{\frac{1}{LC}}$$

Although these models are interchangeable, the waveguide model is usually more useful for transmission line analysis. For the remainder of the chapter, I will focus on the waveguide model.

REFLECTIONS

Whenever an electromagnetic wave encounters a change in impedance, some of the signal is transmitted and some of the signal is reflected (Figure 8.5). The interface between two regions of different impedances is an impedance boundary. An analogy helps in understanding this

Figure 8.5 At a boundary between two regions of different impedance, Z_1 and Z_2, some of the incident energy passes through the boundary, and some is reflected.

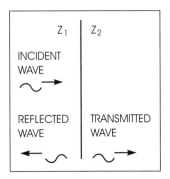

concept. Imagine yourself sitting in a small boat on a pond, looking down into the water. A fish swimming by sees you from below the water's surface because water is transparent. In addition, you faintly see your image as a reflection on the water's surface. Hence, some of the light from your image travels through the water to the fish, and some of the light reflects back to you. This phenomenon occurs because of the difference in optical impedance of water and air.

This same phenomenon also occurs at electronic-signal frequencies. The difference between the impedances determines the amplitude of the reflected and transmitted waves. The reflection coefficient, ρ_V, for the voltage wave is

$$\rho_v = \frac{V_{reflected}}{V_{incident}} = \frac{Z_L - Z_o}{Z_L + Z_o}$$

whereas the transmission coefficient is

$$T_v = \frac{V_{transmitted}}{V_{incident}} = 1 + \rho_V$$

The formulas for the reflection and transmission coefficients of the current wave are slightly different from the voltage wave. The reflection coefficient, ρ_I, for the current wave is

$$\rho_I = \frac{I_{reflected}}{I_{incident}} = -\rho_V$$

and the transmission coefficient is

$$T_I = \frac{I_{transmitted}}{I_{incident}} = 1 + \rho_I$$

The total power reflected is

$$R_P = \frac{P_{reflected}}{P_{incident}} = |\rho_V|^2$$

and the total power transmitted is

$$T_p = \frac{P_{transmitted}}{P_{incident}} = 1 - R$$

The power transmitted added to the power reflected is equal to the incident power, as required by the law of conservation of energy.

An interesting and often underemphasized fact is that the amount of reflection is independent of frequency and occurs at all frequencies (as long as the materials have an approximately constant permittivity over the frequency range of interest). This fact seems contrary to the common belief that reflections are high-frequency phenomena. It is not that reflections don't happen at low frequencies. Even at audio frequencies, this process of reflection occurs. Reflections are just not typically noticed at low frequency. The next section should make clear the reasons why reflections are typically noticed only at high frequency.

PUTTING IT ALL TOGETHER

In summary:

- A signal traveling along a transmission line has voltage and current waves related by the characteristic impedance of the line.
- Signal reflections occur at impedance boundaries.
- As it travels down the line, a signal has delay associated with it.

These three elements combine to produce transmission line effects. The first two items imply that a circuit has reflections unless the

Figure 8.6 In this transmission line example, a source and load are connected through a 1 m Teflon coaxial cable with a characteristic impedance of 50 Ω. Reflections occur at the boundary between the source and the cable because of the difference in impedance. Reflections also occur at the boundary between the load and cable. Assume that the source-to-coax wires and load-to-coax wire are of negligible length.

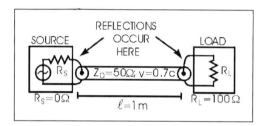

transmission line, source, and load impedances are all equal. The third item implies that reflected waves reach the load staggered in time. An example helps to illustrate. Figure 8.6 shows a transmission line (coaxial cable) with a source and load connected.

Let the source produce a 1 V step function at time $t = 0$. This wave travels down the cable and reaches the load at time $\tau = l/v = 1/(0.7c)$ = 5 nsec. Because the cable's characteristic impedance (50 Ω) is different from the impedance of the load (100 Ω), some of the incident wave transmits to the load, and some is reflected by the load. The reflected wave travels back through the cable and arrives back at the source at time 2τ. Because the source impedance (zero) does not match the characteristic impedance of the cable (50 Ω), another reflection occurs. This reflection travels toward the load and arrives at the load at time 3τ. As with the initial wave, the load absorbs some of the wave, and some is reflected. Subsequent waves arrive at the load in this manner ad infinitum, decreasing in amplitude after each round trip. The net effect is that the load receives a voltage signal that is the superposition of the initial incident wave and all of the subsequent waves (Figure 8.7). As this example demonstrates, the resulting signal at the load can look much different from the original source signal. A simple step signal at the source ends up producing a step wave followed by a series of oscillations at the load. These oscillations eventually settle to the value you would expect if you were to ignore the transmission line.

Figure 8.7 This lattice diagram shows the propagation of a 1 V step wave and its subsequent reflections along the transmission line of Figure 8.6. The waves received at the load are shown at the right of the diagram. The diagram shows the first five reflections.

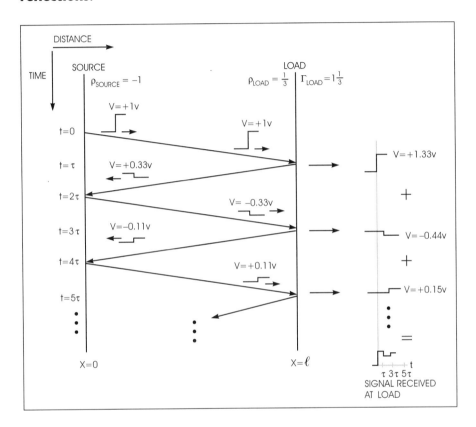

DIGITAL SIGNALS AND THE EFFECTS OF RISE TIME

In this example, the step signal is an ideal digital wave; that is, a signal with zero rise time. Of course, real-world step signals have rise times greater than zero. Changing the rise time of the step signal changes the shape of the signal that appears at the load. Figure 8.8 illustrates the load waveforms for the same transmission line using various rise times. When the rise time becomes much longer than the transmission line delay (τ), the reflections get "lost" in the transition region. The effect of the reflections then becomes negligible. It is important to note that, regardless of the rise time, the amplitude of the reflections is the same. The rise time affects only the superposition of the reflections. The transmission line

Figure 8.8 This figure illustrates the dependence of transmission line side-effects on signal rise time. In each case, (A), (B), and (C), a 1V step source signal is applied to the circuit of Figure 8.6. The reflections are of the same amplitude in each case, but the resulting load signals are quite different. (The diagrams show only the first four waves to read the load.)

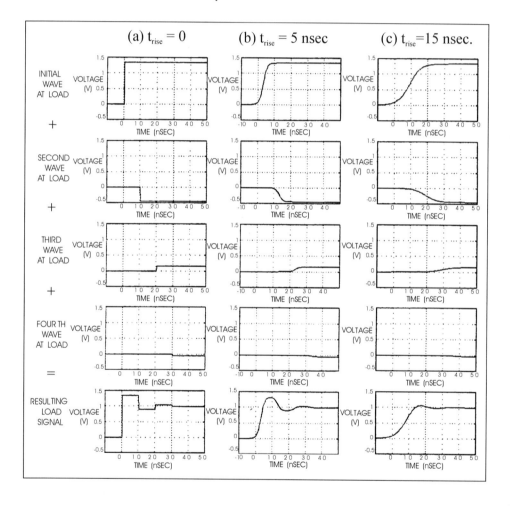

effects (over-shoot and oscillation) become apparent when the rise time, t_{RISE}, is short compared with the transmission line delay, τ. Such signals are therefore in the domain of high-frequency design. When t_{RISE} is long compared with τ, the transmission line effects are negligible; these signals are in the domain of low-frequency design. *For most applications, you can consider* $t_{RISE} > 6\tau$ *to be in the low-frequency domain.*

ANALOG SIGNALS AND THE EFFECTS OF FREQUENCY

You can use a similar analysis for analog signals. Using the circuit of Figure 8.6, let the source be a 1 V sine wave with a frequency of 1 MHz (period $T = 1\,\mu sec$). This signal undergoes the same reflections as the step signal. The amplitude of the waves incident on the load is also the same, namely +1.33 V, −0.44, +0.15, −0.05, etc. The sum of the series of waves is a 1 V sine wave (Figure 8.9B). In this example, no transmission line effects are noticeable because the delay of the transmission line is negligible. In fact, the round-trip cable delay is 1/100th the signal period. In addition to the voltage wave, there is also a current wave at the source and load. For the 1 MHz signal, the current wave is approximately the same at the source and at the load. At both the source and the load, the current wave is a 10 mA (1 V/100 Ω) sine wave that is exactly in phase with the voltage wave.

In contrast, if you increase the source frequency to 50 MHz (period $T = 20\,nsec$), the resulting load signal is quite different, even though the incident waves are of the same amplitude. The delay through the transmission line is no longer negligible at this higher frequency (Figure 8.9A). In fact, the round-trip delay of the transmission line causes the incident waves to shift 180°. With these phase shifts, the incident waves produce a load signal that gradually builds to a steady-state 2 V sine wave! In addition, the load signal shifts in phase from the source signal. The current wave will also be affected at this frequency. At the load, the resulting steady-state current is governed by Ohm's law. Therefore the load current wave is simply a 20 mA (2 V/100 Ω) sine wave, which is exactly in phase with the voltage wave. At the source, however, the current wave is a 40 mA sine wave. The power at both locations is the same, $P = I \times V = 40\,mW$, but the voltage and current are different. Such a transmission line acts like a 2:1 transformer. At the source end, the impedance of the load appears as only 25 Ω. In fact, this circuit, a transmission line that is exactly one-quarter wavelength long, is often called a quarter-wave transformer.

For this circuit, 50 MHz falls in the domain of high-frequency design, and 1 MHz falls in the domain of low-frequency design. For most applications, you can consider a sine-wave period of $T = 20\tau$ (or $\lambda = 20L$) the boundary between the high- and low-frequency domains.

To avoid problems from reflections, RF systems are designed around a standardized characteristic impedance. Typically $Z_o = 50\,\Omega$ is used (as shown in Figure 8.10). This is why most lab function generators and test equipment have a source impedance of 50 Ω. Young engineers often learn this fact the hard way (as I did); they turn on a function

Figure 8.9 This figure illustrates the dependence of transmission line side-effects on signal frequency. In each case, a 1 V sine wave signal is applied to the circuit of Figure 8.6. The reflections are of the same amplitude in each case, but the resulting load signals are quite different. Notice the relative phase between the reflections and the original signal in each case.

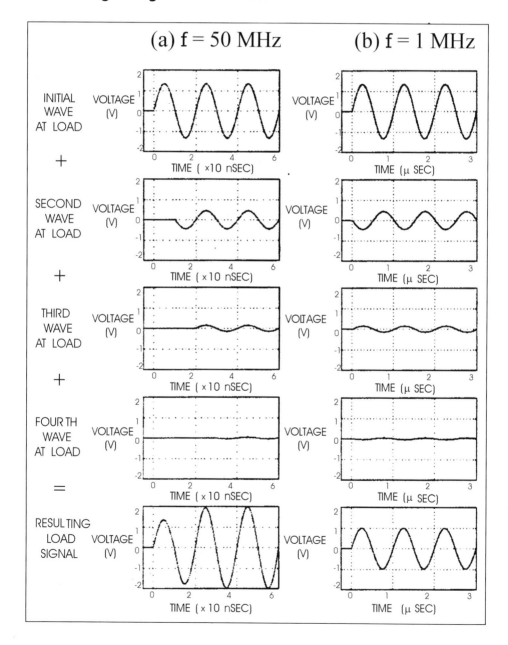

IMPEDANCE TRANSFORMING PROPERTIES   167

Figure 8.10 If a transmission line is matched on both ends, wave reflections will not occur. Most test equipment and RF systems are designed for cables with a characteristic impedance of 50 ohms.

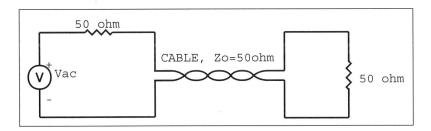

generator in the lab and set it to 1 volt. The engineer measures the open circuit output voltage to be 2 volts instead of 1 volt. The simple answer to this problem is that the voltage will only be 1 volt if a 50 ohm load is placed across the output. The function generator has a 50 ohm source impedance, so that it matches the impedance of typical coaxial cable used for test equipment.

IMPEDANCE TRANSFORMING PROPERTIES

As the previous circuit demonstrated, transmission lines transform the impedance of loads. For transmission lines that are electrically short ($L \ll \lambda$) the effect is negligible, but for transmission lines that are electrically long the effects can be dramatic. It is important to remember that the behavior of a transmission line changes every quarter wavelength.

Figure 8.11 shows that when the length is less than a quarter wavelength, an open circuit appears as a capacitance and a short circuit appears as an inductance. In more general terms, when the load impedance is greater than the characteristic impedance of the cable, a capacitance occurs in parallel with the load. When the load impedance is less than the characteristic impedance of the cable, an inductance occurs in series with the load. Of course, if the load impedance is equal to the characteristic impedance of the cable, the load appears exactly as it is. In digital systems, the input impedance of most devices is typically quite high. Therefore, the designer must take into account the fact that the transmission line will likely present a capacitance to the device driving the signal.

Figure 8.11 These circuits show the equivalent impedance for a short transmission line terminated in various loads. The source is matched, and thus the voltage at the source end of the cable is superimposed with a single reflection. The open-circuit load (A) appears as a capacitance and the short-circuit load (B) appears as an inductance. Circuits (C) and (D) show the results of a short transmission line terminated with a resistor.

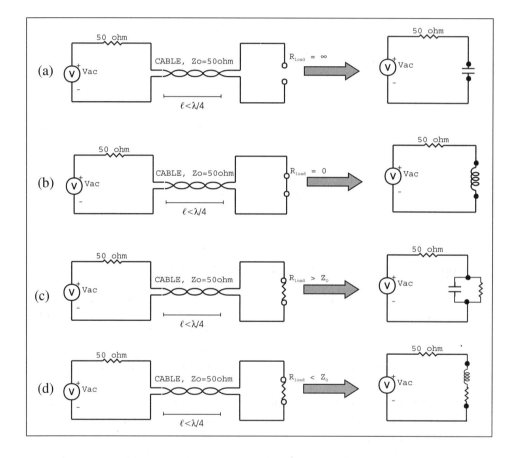

As a transmission line's length is increased from zero to $\lambda/4$, the reactance increases until the length reaches exactly $\lambda/4$. For an open-circuited line, the capacitance increases until, at exactly $\lambda/4$, the capacitance is infinite. In other words, the input impedance of the open-circuited transmission line becomes a short circuit! A short circuit, in turn, appears as an open circuit. As illustrated in Figure 8.12, these effects are resonant conditions that occur only when a quarter-

IMPEDANCE TRANSFORMING PROPERTIES 169

Figure 8.12 These circuits show the equivalent impedance for a quarter-wavelength transmission line terminated in various loads. The source is matched, and thus the voltage at the source end of the cable is superimposed with a single reflection. Because the reflection travels exactly a half-wavelength, the reflection will be 180 degrees out of phase with the source wave. The open-circuit load (A) appears as a short. The short-circuit load (B) appears as an open. Circuit (C) shows the results of a quarter-wavelength transmission line terminated with a resistor.

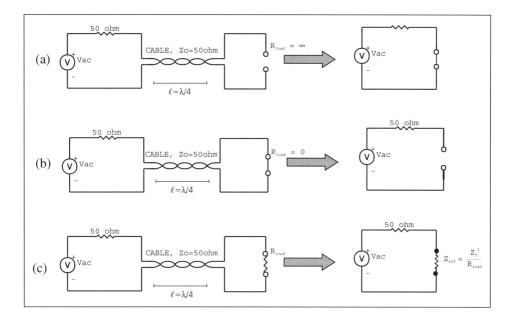

wavelength of the signal is equal to the length of the line. For this reason, open-circuited or short-circuited quarter-wavelength transmission lines can be used as resonators for oscillators and filtering applications. If the load is not a short-circuit or open circuit, the transmission line acts as a transformer, transforming the load impedance to an effective impedance of

$$Z_{effective} = \frac{Z_0^2}{R_{Load}}$$

As the transmission line length is increased above a quarter-wavelength, the roles reverse, in a manner of speaking. The open circuit

Figure 8.13 Equivalent impedances for a transmission line of intermediate length ($\lambda/4 < \ell < \lambda/2$) terminated in various loads. The source is matched, and thus the voltage at the source end of the cable is superimposed with a single reflection. The resulting impedances are opposite to those of the short transmission line circuits. The open-circuit load (A) appears as an inductance, and the short-circuit load (B) appears as an capacitance. Circuits (C) and (D) show the results of the transmission line terminated with a resistor.

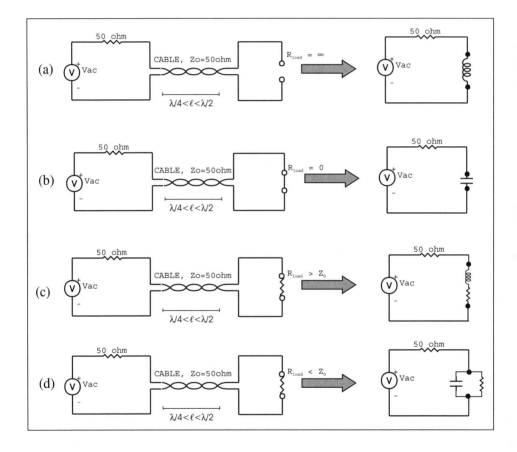

appears as an inductance, and the short-circuit appears as a capacitance. The effective impedance of intermediate-length transmission lines terminated in various loads is shown in Figure 8.13. The reactance increases as the length is increased, until when the length is exactly $\lambda/2$, the open circuit appears as an open circuit and the short-circuit appears as a short-circuit. At $\lambda/2$, all load resistances appear as their exact value. As length is increased from this point, the entire pattern repeats itself and continues forever following the same pattern through every half wavelength.

IMPEDANCE MATCHING FOR DIGITAL SYSTEMS

In digital systems, transmission line reflections can cause oscillations (ringing), overshoot/undershoot, and "shelves." Overshoot/undershoot appears whenever the load resistance, is greater than the characteristic impedance of the transmission line. A shelf or "non-monotonicity," is a reflection phenomenon in which the load voltage temporarily remains at an intermediate voltage. For example, in a 5 V logic system, a shelf can appear that is at around 2 volts. The shelf appears when the initial wave reaches the load and then eventually goes away as the subsequent reflections arrive at the load. Shelves should be avoided at all costs, because during this period the voltage can be in the transition region of the load device. Small amounts of noise can cause the device to produce glitches. Shelves only occur when the load impedance is lower than the transmission line characteristic impedance.

To avoid these phenomena, the load can be matched to the characteristic impedance of the transmission line by placing a resistor in parallel to the load. The overall resistance at the load should never be less than the characteristic impedance to ensure that shelves do not occur. A suitable alternative in many cases is to match the source device to the line impedance. Source matching is accomplished by adding a series resistance to the output of the source device such that this resistance when added to the internal source resistance is approximately equal to the characteristic impedance of the transmission line. Keep in mind that source matching deceases the output voltage of the signal and thereby increases its susceptibility to noise. Matching both the load and source is the ideal cure for eliminating reflections, but in most cases is unnecessary.

There are several variations on these basic techniques. One power-saving technique is to use a combination of a resistor and a capacitor at the load so that the combination produces a matching impedance during the transitions, but forms an open circuit during periods where the signal is at the same level for a long period of time. Another technique combines impedance matching with data bus pull-up resistors. With this technique, one resistor is placed from the load input to ground, and a second resistor is placed from the load input to supply voltage. When no devices are driving the data bus, these resistors form a voltage divider, keeping the voltage at a defined level. When the data bus is in operation, the two resistors provide a parallel resistance to match the characteristic impedance for AC signals. Once you understand the basics, these other techniques are easy to add to your repertoire. Often the best place to look for ideas on matching is the application notes of the IC manufacturers themselves. Texas Instruments and Fairchild Semiconductor are two companies that have many useful digital designer application notes on their web sites, covering digital system impedance matching and other analog design considerations for digital devices.

An important footnote to this section is the design of the transmission lines themselves. On printed circuit boards (PCBs), designers use either microstrip (a copper trace above a ground plane) or strip-line (a copper trace between two ground planes). The characteristic impedance of such transmission lines is dependent on three parameters: 1) the dielectric constant of the PCB material, 2) the width of the trace, and 3) the height of the trace above the ground plane (i.e., the thickness of the PCB layer). The formulas are quite complex, and can be found in many books, such as Pozar (1998).

Typically, the trace width is determined by space constraints and is often made very narrow to increase the component density. The thickness of the PCB layer is usually determined to a large extent by how many PCB layers are used. The digital designer should be aware of what these parameters will be to facilitate determining a good estimate of the characteristic impedance during the design phase and not leave such calculations to the last minute. Also keep in mind that most PCB manufacturers only specify the layer thickness in a very loose manner. If you want better precision for thickness, you will likely have to notify the manufacturer upfront and pay higher manufacturing costs.

IMPEDANCE MATCHING FOR RF SYSTEMS

Impedance matching is even more important in RF systems, and the techniques are slightly different. For RF transmission lines, the main goal is the same as for digital transmission lines: eliminate reflections. The first reason for eliminating reflections in RF systems is to assure that proper voltage is transferred between the source and load. As the earlier example showed, transmission line reflections can cause the load voltage to be different than the voltage on the source end. Without properly matching the load to the transmission line, the voltage is dependent on the length of the transmission line. The second reason to eliminate reflections is to present a known impedance to the source. If the load is not matched, the impedance at the source may be very small or very big. A very small impedance will overload the source amplifier, causing undesired behavior and possibly damage to the source amp. Reflections can also create reactive impedances (capacitance or inductance) at the source. Many amplifiers will not be stable and will oscillate if a large reactive load is connected to the output. The third reason for matching the load to the transmission line is to reduce losses in the transmission line itself. In many cases, a transmission line can be approximated as lossless, but there are notable exceptions, the two most common being where lost power is very costly or where the lines are long (as loss is proportional to length). For antennas both of the conditions apply. Transmitting antennas often require high power, and any power lost to heating in the transmission line cable is money wasted. For receiving antennas, bringing the largest signal to the amplifier is paramount; any loss in the cable diminishes signal strength. Antennas are also often considerable distances from the transmitting or receiving amplifier. Reflections imply that energy is flowing back and forth along the line. At any given moment, the sine wave output at the load is a superposition of the initial wave and the subsequent reflections. Each wave that flows along the transmission line will lose a certain percentage of energy with each trip down the line. In an unmatched system, much of the energy must make one or more round-trips down the line before being dissipated in the load (or transmitted on an antenna). Refer back to Figure 8.9A. In this example, only about two-thirds of the load energy arrives via the initial wave. The rest of the load energy arrives in the subsequent reflections. Therefore, matching the load and source to the transmission line reduces power loss in the transmission line.

MAXIMUM LOAD POWER

Another reason for impedance matching in RF systems is to deliver the maximum power to the load. The law of maximum power transfer states that given a voltage source with internal resistance, R_s, the maximum power is transferred to the load when the load resistance is equal to the source resistance. Furthermore, the reactive impedance of source and load should be such that each cancels out the other. Stated mathematically, you should set $Z_L = R_s - (i \times X_s)$, where X_s is the reactance (imaginary impedance) of the source. The principle is important in all aspects of RF systems design. To transfer the maximum power from stage to stage inside an RF design, the output impedance of one stage should be matched to the input impedance of the next stage. This rule is important not only for amplifiers and transmission lines, but also for passive stages such as filters. RF filters should be designed such that the filter inputs and outputs match the connected stages for signals inside the passband. For signals outside the passband, matching is not needed. However, be sure to design, simulate, and test the filter with the impedances of the previous and following stages connected. In other words, if your filter output connects to a 50Ω cable, be sure to include a 50Ω load as a design parameter. The filter will behave much differently with the output open-circuited.

Power Efficiency

You may wonder why power utilities don't follow the law of maximum power transfer. There is a very good reason for this; the power company has a much different goal. The power company's goal is to transmit a standard voltage (a nominal 120 volts RMS in the U.S.) to its customers in the most efficient manner possible. They are not interested in delivering the maximum power to our houses. If they did, the voltage would increase and blow out all the electrical appliances in our houses. Furthermore, the law of maximum power transfer is not the same as the law of most efficient power transfer. To transfer power most efficiently from source to load, the ratio of load resistance to source resistance should be made as large as possible. One way this can be accomplished is by lowering the source resistance as much as possible.

To reduce the power lost in the power transmission line, the source voltage is stepped up using transformers. By increasing the voltage, the same amount of power can be transmitted down the line but with a much smaller current. Since most of the losses in a transmission line are in the wires themselves (as opposed to losses in the dielectric between the wires), reducing the current reduces the most transmission line loss. (Power lost in a wire is equal to I^2/R_{wire}.) It is for this reason that electric utilities use transformers to deliver power at voltages as high as approximately 1 megavolt.

Why 50 Ohms?

A question you may have is: Why is 50 ohms the most common impedance chosen for RF systems? In Chapter 7, you learned that large resistances are susceptible to capacitive parasitics, and small resistances are susceptible to inductive parasitics. The characteristic impedance should thus be an intermediate value, so that filter and amplifier impedances can be reasonably valued. Furthermore, making the impedance very small would place difficult constraints on the amplifier design. For example, if the impedance were 1 ohm, you would have a very difficult time designing amplifiers since the input and output impedance would need to be 1 ohm.

But why 50? The answer is actually quite simple, and it derives from the characteristics of coaxial cables. The geometry of coaxial cables is such that an air-filled cable designed for maximum power handling (i.e., designed such that the breakdown voltage of the air is highest) has a characteristic impedance of 30 ohms. An air-filled coaxial cable designed for maximum power efficiency (least loss) has an impedance of 77 ohms. Therefore 50 ohms serves as a good compromise between the two optimal impedances for air-filled dielectrics. Furthermore, for typical solid cable dielectrics such as Teflon, the impedance for least loss is about 50 ohms. It follows that typical 50 ohm cables are optimized for least loss. A further reason for 50 ohms is that monopole antennas with ground radials have a characteristic impedance of about 50 ohms, making 50 ohm coaxial cable a good match to monopole antennas.

What about 75 ohm cable? If you want to design a cable for lowest loss, you should choose air for a dielectric because its loss is lower than any solid material's. From the previous paragraph, you know that 77 ohms is the optimum impedance for air-filled cable. An air-filled dielectric with an impedance of around 77 ohms is therefore the most power-efficient cable possible. Cable TV companies standardized on 75 ohms because their cables run over very long distances and therefore loss is an important design factor. Telephone companies also often use 75 ohm cables for their interoffice "trunk" lines, which carry multiplexed telephone signals. To make an air-filled cable requires the use of dielectric spacers at regular intervals. Trilogy Communications (www.trilogycoax.com) makes such cables specifically for the CATV market.

MEASURING CHARACTERISTIC IMPEDANCE: TDRS

Although formulas are readily available for calculating the characteristic impedance of most transmission line geometries, no formula can ever give the assurance of physical measurement. I encourage all designers, RF and digital, to measure the typical characteristic impedance of the transmission lines in their designs. There are several ways to go about measuring characteristic impedance. The most useful method for measuring the parameters of transmission lines is to use a time domain reflectometer (TDR). The name is intimidating, but the equipment is rather easy to use. A TDR is basically an oscilloscope with a built-in step generator. The step generator sends out a voltage step with a very fast rise time, approximating the signal of Figure 8.8A. You then watch the screen and measure the amplitude of the reflected waves. If all ends are matched, the voltage is a simple straight line. If there are mismatches, the TDR or you can calculate the impedances of the mismatches. The timing of the reflections allows you to calculate the speed of light along the transmission line if you measure physical length, $v = 2L/\Delta t$ (the 2 in the numerator is required because the reflection must make a round-trip down the line). You can then use the velocity to calculate the dielectric constant of dielectric material,

$$\varepsilon = \frac{1}{\mu v^2}$$

The TDR works in much the same way radar operates in the open air, except that radar uses impulses instead of step functions. A TDR can also give you information about discontinuities or defects along the line, because any deviation of the characteristic impedance will cause a reflection. Basically, the time domain signal on the screen gives you a picture of the impedance as a function of distance along the cable. This technique is very useful for finding shorts or insulation cracks in long computer network, CATV, or phone line cables, especially if the cable is underground or in a wall. You can often purchase TDR modules for high-end oscilloscopes. Another option is to create your own using a combination of an oscilloscope and a function generator, which can generate step functions with very fast rise times. The rise time determines the distance resolution you can achieve. For example, a 1 nanosecond rise time will typically be sufficient for 1 meter of resolution.

A simpler method for determining characteristic impedance involves direct measurement of the impedance. This technique works quite well, but doesn't provide the ability to locate discontinuities and faults. You measure complex impedance of the transmission line with the load end short-circuited and then perform the same measurement with the load end open-circuited. The transmission line characteristic impedance is calculated using the formula,

$$Z_o = \sqrt{Z_{short} \times Z_{open}}$$

This technique requires that you have a method to measure complex impedances. An impedance analyzer, a vector network analyzer, or a vector voltmeter can be used for this task. Although transmission line characteristic impedance is usually quite constant across frequency, it can be very different at very high and very low frequencies. It is therefore important that your measurements are made with a signal whose frequency is in the approximate range of the signal frequency of your system. For example, a typical 26 gauge twisted-pair telephone line will have a characteristic impedance of $Z_o \cong 90\,\Omega$ above 10 MHz, but will have a complex, lossy characteristic impedance in audio frequencies (e.g., $Z_o \cong 650 - [i \times 650\,\Omega]$ at 1 kHz).

STANDING WAVES

The final topic of this chapter on transmission lines is standing waves. On a matched transmission line, there are no reflections. The voltage and current waves at the load and source end are the result of a single traveling wave. Furthermore, the voltage and current at any point along the line are the result of a single wave. If you measure the voltage and current at any point along the line, you will observe two synchronized sine waves and the magnitude ratio of the two will be exactly $V/I = Z_o$. Standing waves do not occur on matched transmission lines.

With an unmatched transmission line, the observations will be much different. As you learned earlier, the voltage and current waves at the load and source are the superposition of the initial wave and an infinite series of reflections. A matched transmission line has energy flowing in one direction, from source to load. In contrast, an unmatched transmission line has an infinite number of waves reflecting back and forth along the line at all times. This fact is very important to remember. Remembering this will keep you out of trouble when faced with transmission line problems. The characteristic impedance relation $V/I = Z_o$ holds for each individual wave at any intermediate point along the transmission line. However, except when a signal is first applied, every point along the line will exhibit a superposition of all the reflections. Therefore, the observed signals, which are a superposition of all the waves, will not be governed by the characteristic impedance ($V/I \neq Z_o$ for unmatched lines).

A result of the reflections is that a standing wave is produced on the line. The standing wave takes the form of a sine wave with a wavelength equal to the signal wavelength. This standing wave determines the amplitude of the signal at each location. In other words, the signal is amplitude-modulated in a spatial sense. At every location, the signal varies in time as a sine wave, but the amplitude of the sine wave is governed by the standing wave. Furthermore, the standing wave for the current and the standing wave for the voltage have the same wavelength (furthermore, $\lambda_{\text{standing-wave}} = \lambda_{\text{signal}}$), but not the same phase. The voltage and current standing wave are exactly 90 degrees (1/4 λ) out of phase. Spaced every quarter wavelength along the line is a node. At each node either the voltage or current standing wave will be a minimum. At these points the resulting impedance is maximum or minimum for the respective cases. This situation may seem reminiscent of the earlier section on the impedance transformation aspects of transmission lines. You learned that the input impedance of a transmission line alternates between a maximum and minimum every quarter-wavelength. In fact, this is the same phenomenon, viewed from a different perspective. The wave impedance at distances along the transmission line varies in the same way the transmission line input impedance varies when the line length is changed. At the standing wave nodes, the wave impedance is $Z_{node} = Z_L$ or $Z_{node} = Z_o^2/Z_L$. Notice that when the load is matched, $Z_{node} = Z_L = Z_o$, the impedance takes on intermediate values at locations in between the nodes.

Infinity Becomes Two

For sinusoidal signals, the steady-state behavior can be described by combining all the waves into a single forward traveling wave (i.e., traveling toward the load) and a single backward traveling wave (i.e., traveling toward the source). This steady-state analysis is used almost exclusively for analog systems because it simplifies matters greatly. On the other hand, steady-state analysis is much less useful in digital systems since it only predicts the steady-state voltage and not the transient phenomena like ringing.

BIBLIOGRAPHY

Bogatin, E., "What Is Characteristic Impedance," *Printed Circuit Design*, January 2000.
Bowick, C., *RF Circuit Design*, Carmel, Indiana: SAMS Publishing, 1995.
Carr, J. J., *Practical Antenna Handbook*, 4th Edition, New York: McGraw-Hill, 2001.
Chan, K. C., and A. Harter, "Impedance Matching and the Smith Chart—The Fundamentals," *RF Design*, July 2000.
Gupta, K. C., R. Garg, I. Bahl, P. Bhartia, *Microstrip Lines and Slotlines*, 2nd Edition, Boston: Artech House, 1996.
Hewlett-Packard, *Time Domain Reflectometry Theory*, Hewlett-Packard, AN-1304-2.

BIBLIOGRAPHY

Johnson, H., "Differential Termination," *EDN*, June 5, 2000.
Johnson, H., "Why 50 Ω?," *EDN*, September 14, 2000.
Johnson, H., "50 Ω Mailbag," *EDN*, January 4, 2001.
Johnson, H., "Both Ends of Termination," *EDN*, January 18, 2001.
Johnson, H., "Strange Microstrip Modes," *EDN*, April 26, 2001.
Johnson, H., and M. Graham, *High-Speed Digital Design: A Handbook of Black Magic*, Englewood Cliffs, NJ: Prentice-Hall, 1993.
Jordan, E. C., and K. G. Balmain, *Electromagnetic Waves and Radiating Systems*, 2nd Edition, Englewood Cliffs, NJ: Prentice Hall, 1968.
Kraus, J. D., and D. A. Fleisch, *Electromagnetics with Applications*, 5th Edition, Boston: McGraw-Hill, 1999.
Matthaei, G., E. M. T. Jones, L. Young, *Microwave Filters, Impedance-Matching Networks, and Coupling Structures*, Boston: Artech House, 1980.
Montrose, M. I., *Printed Circuit Board Design Techniques for EMC Compliance—A Handbook for Designers*, 2nd Edition, New York: IEEE Press, 2000.
Paul, C., and S. Nasar, *Introduction to Electromagnetic Fields*, 2nd Edition, Boston: McGraw-Hill, 1987.
Pozar, D. M., *Microwave Engineering*, 2nd Edition, New York: John Wiley, 1998.
Ramo, S., J. R. Whinnery, T. Van Duzer, *Fields and Waves in Communication Electronics*, 2nd Edition, New York: John Wiley, 1989.
Rosenstark, S., *Transmission Lines in Computer Engineering*, Boston: McGraw-Hill, 1994.
Schmitt, R., "Analyze Transmission Lines with (almost) No Math," *EDN*, March 18, 1999.
Straw, R. D., Editor, *The ARRL Antenna Book*, 19th Edition, Newington, Conn.: American Radio Relay League, 2000.
Sutherland, J., "As Edge Speeds Increase, Wires Become Transmission Lines," *EDN*, October 14, 1999.
Wadell, B. C., *Transmission Line Design Handbook*, Boston: Artech House, 1991.

Web Resources

Some great animated tutorials on transmission lines can be found at:
http://www.williamson-labs.com/xmission.htm
http://www.elmag5.com/programs.htm
http://hibp.ecse.rpi.edu/~crowley/javamain.htm
http://hibp.ecse.rpi.edu/~crowley/java/SimWave/simwav_g.htm

9 WAVEGUIDES AND SHIELDS

Waveguides and shields may seem like an odd combination for a chapter, but the two actually have a lot in common. The operation of both devices depends on the reflectivity of metals. Metals are highly reflective through most of the electromagnetic spectrum. They reflect radio waves as well as visible light. Although metals have different properties at visible light frequencies, and aren't conductors in this band, they still reflect most of the incident radiation if the surface is well polished and not tarnished. We take advantage of this whenever we look in the mirror. A mirror is just a piece of glass with a metal film deposited on the backside. The glass serves to protect the metal surface from tarnishing and provides a smooth substrate on which to deposit the metal. Because glass also has a higher dielectric constant than air, the glass also increases the metal's reflectivity.

You are also probably aware that metal reflects radio waves. This is why radar works so well. The microwave radiation bounces off the car or airplane and returns to the radar dish. Radio waves can also be reflected by the ground, which acts as a decent conductor in many radio bands.

The simplest electromagnetic shield is just a conductor sheet. To learn how a shield works, assume for the moment that it is made from a perfect conductor. A perfect conductor is a conductor with zero resistance. As a consequence, the electric field must be zero inside a perfect conductor, otherwise the current would become infinite. For perfect conductors, all current must travel in an infinitesimally thin layer at the surface of the conductor. Since an electric field cannot exist inside a perfect conductor, it follows that electromagnetic waves cannot travel into a perfect conductor. When an electromagnetic wave impinges on a perfect conductor, 100% of the wave energy is reflected because no energy can enter the conductor, thus explaining that a perfect conductor is also a perfect reflector.

REFLECTION OF RADIATION AT MATERIAL BOUNDARIES

Wherever a boundary between two materials occurs, reflections of waves will occur. The situation is just like that of impedance boundaries in transmission lines. The reflected waves can be calculated from the electric and magnetic reflection equations:

$$\rho_E = \frac{\eta_2 - \eta_1}{\eta_2 + \eta_1}$$

$$\rho_H = \frac{\eta_1 - \eta_2}{\eta_1 + \eta_2}$$

where η_1 is the *intrinsic material impedance* of the first region and η_2 is the intrinsic material impedance of the second region. The electric field is analogous to the voltage in a transmission line, and the magnetic field is analogous to the current in a transmission line. For air the intrinsic impedance is approximately that of free space:

$$\eta_0 = \sqrt{\frac{\mu_0}{\varepsilon_0}} \cong 377\,\Omega$$

For a conductor, the intrinsic impedance is

$$\eta = \sqrt{\frac{\mu}{\varepsilon - i\frac{\sigma}{2\pi f}}}$$

Figure 9.1 plots the intrinsic impedance of copper across a wide frequency range. When a radio wave traveling through air encounters a slab of copper, a reflection results. The resulting reflection coefficient is complex, implying a phase shift of the reflected wave.

Figure 9.2 plots the reflection coefficient for a radiating wave incident upon a thick sheet of copper for a broad range of frequencies. Notice that the vast majority of the incident power is reflected at all frequencies. This result is valid only for radiating or far-field electromagnetic energy. As you learned in Chapter 5, the wave impedance (ratio of electric field to magnetic field) in the far field is equal to the intrinsic impedance of the medium in which the wave is traveling. This statement is not true in the near field.

Figure 9.1 Magnitude of wave impedance in copper. In copper, the wave impedance is a complex number and is equally split between real and imaginary parts.

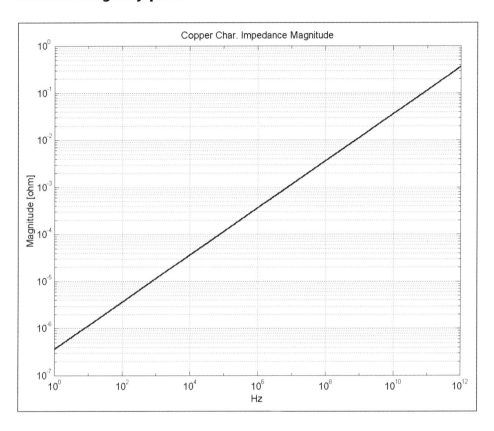

THE SKIN EFFECT

All real conductors exhibit a phenomenon called the *skin effect*. The skin effect gets it name from the fact that any electromagnetic wave incident upon a conductor diminishes exponentially in amplitude as it penetrates the conductor. The skin depth defines the depth at which the amplitude of the wave has diminished to about $e^{-1} \sim 36.9\%$ of its initial value. At a depth of five skin depths, the signal diminishes to about 0.7% of its surface value. Skin depth is inversely proportional to both conductivity and frequency. The exact formula is

$$\delta = \frac{1}{\sqrt{\pi f \mu \sigma}}$$

Figure 9.2 Reflectivity (R) of Copper (plotted as $1 - |R|^2$).

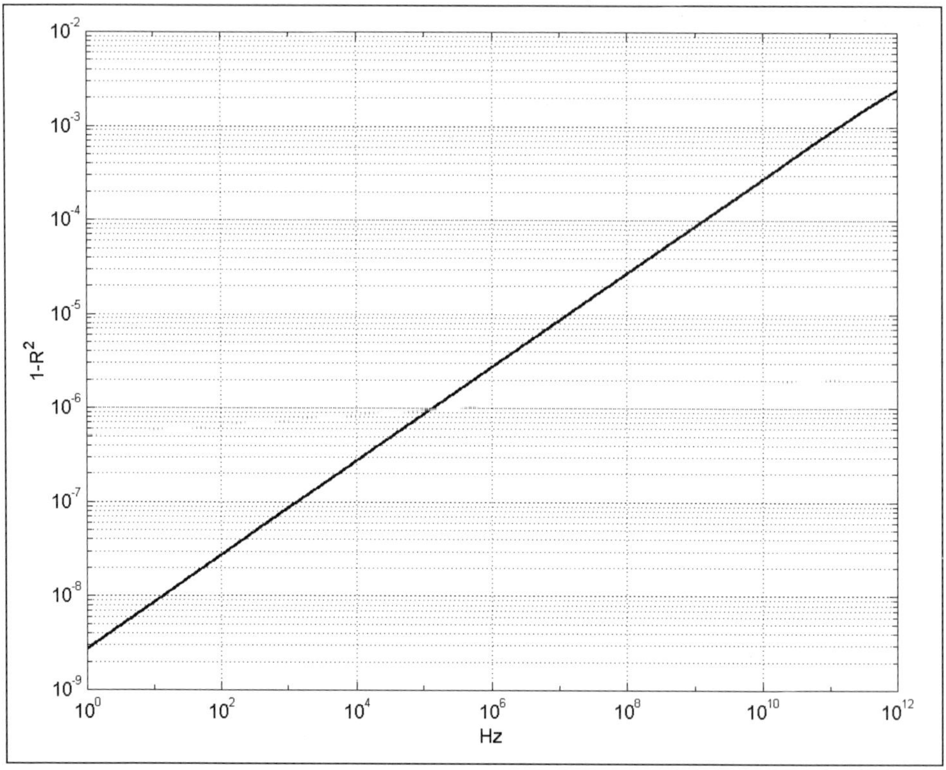

Therefore, as the frequency of the radiation is increased, a conductor becomes more and more like a perfect conductor in the sense that electric fields cannot penetrate it. Figure 9.3 plots the skin depth for copper.

SHIELDING IN THE FAR FIELD

The purpose of a shield is to prevent electromagnetic energy from penetrating through the shield. Conductors provide two mechanisms for shielding. The first shielding mechanism is reflection, which is most important for low-frequency shielding. The second shielding mechanism is absorption, which is caused by the skin effect. Absorption is most important at high frequency. At all frequencies, reflection is responsible for shielding the majority of power. However, it is the absorption caused by the skin effect that makes shielding so dramatically effective at high frequency. Figure 9.4 shows the relative signal power that is reflected,

SHIELDING IN THE FAR FIELD   185

Figure 9.3 Skin depth of copper.

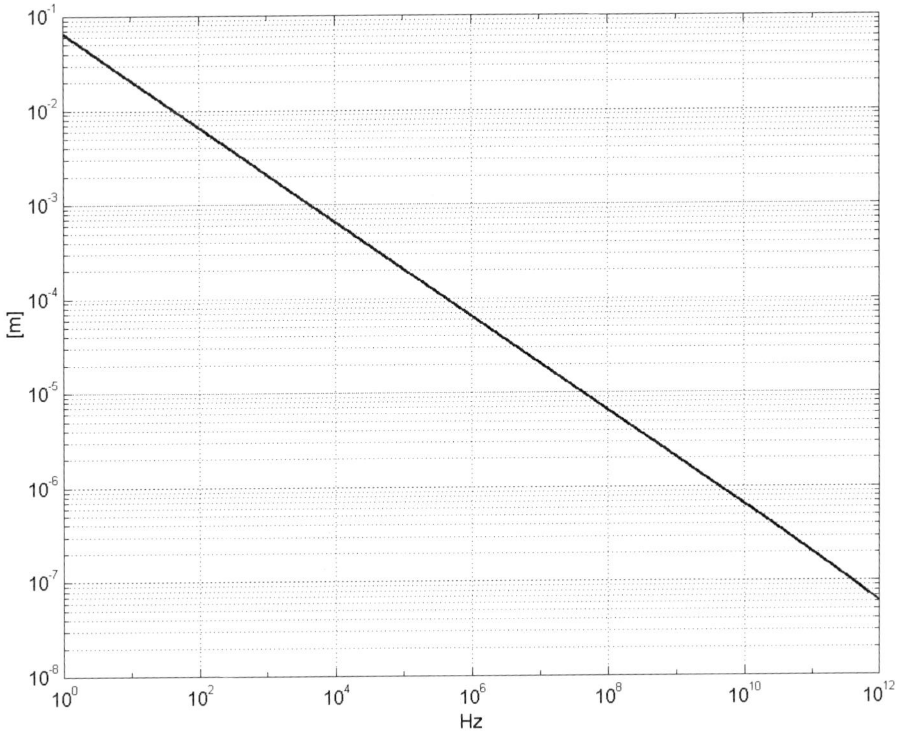

[Graph: Skin depth [m] vs frequency [Hz], log-log plot from 10^0 to 10^{12} Hz on x-axis and 10^{-8} to 10^{-1} m on y-axis, showing a straight decreasing line.]

absorbed, and transmitted for a copper shield. The absolute value of transmitted power is often called the *shielding effectiveness*.

Each quantity in Figure 9.4 is specified in decibels (dB), relative to the power of the incident signal. A power of exactly 0 dB corresponds to a signal of equal power to the incident signal. Although the reflected power is very close to 0 dB, it is never exactly 0 dB. Every drop of 20 dB corresponds to power reduction by a factor of 10. A power level of $-\infty$ dB corresponds to zero power.

Figure 9.5* shows the far field shielding effectiveness for a copper shield, in terms of reflection, absorption, and total shielding effective-

---

*Readers familiar with the shielding literature may notice that the curves in Figure 9.5 are slightly different in shape from those typically seen. The curves in Figure 9.5 were drawn using the exact expressions, whereas approximations are used typically. The correctness of Figure 9.5 was confirmed via personal correspondence with Dr. Clayton Paul.

Figure 9.4 Far-field shielding: reflected, absorbed, and transmitted power for 10 mil thick copper shield.

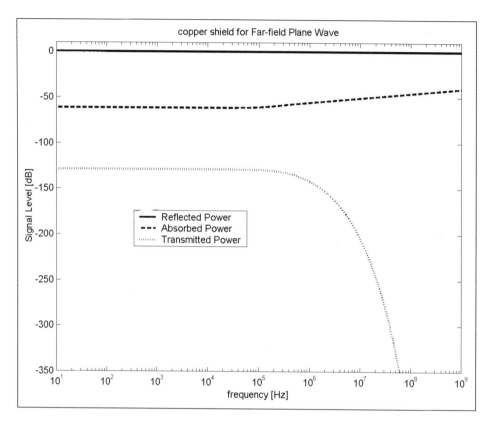

ness. Shielding effectiveness is often used by EMC engineers to quantify how well a shield performs. Shielding effectiveness refers to the ratio of the incident power to the transmitted power, expressed in decibels.

The Effect of Holes in the Shield

If a shield has a hole, some of the incident radiation can avoid the shield and propagate through the hole. The amount of energy that can transmit through the aperture (hole) is greatly dependent on the electrical size of the aperture. If the aperture has dimensions greater than a half wavelength, the aperture lets all power at the opening through. Consider a 1 inch hole in an opaque screen at optical frequencies. Light cast on the opening will pass through unattenuated. The region behind the solid portion of the screen is in shadow. The region directly behind the

Figure 9.5 Far-field shielding effectiveness for 10 mil thick copper shield.

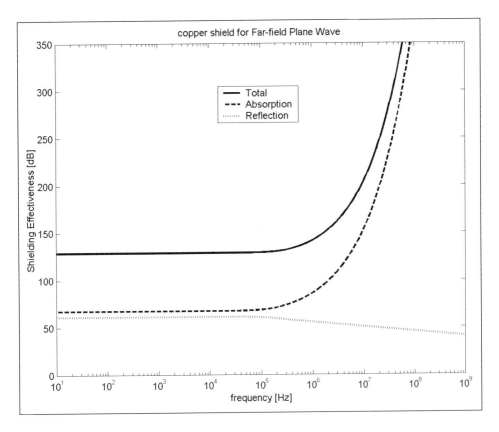

opening is illuminated. The same effect occurs at radio frequencies when the opening is greater than about a half wavelength. The amount of power that passes through the hole is equal to the amount of power incident upon the whole; in other words, the transmitted power is directly proportional to the aperture area.

When the aperture dimensions are reduced below a half wavelength, the radiation is "choked out" due to diffraction and is not proportional to just the area of the aperture. For a circular aperture, the transmitted radiation is mainly dependent on the diameter. For arbitrarily shaped apertures, the transmitted radiation is dependent on the largest dimension of the aperture. If the largest dimension of an aperture is greater than half a wavelength, the aperture will transmit very effectively. Such an aperture is often called a slot antenna.

Thus, for shielding purposes, all apertures in the shield must have dimensions less than half a wavelength. When an aperture is less than half a wavelength in dimension, the transmitted energy dramatically decreases. When the aperture is less than about $1/10\ \lambda$, the transmitted energy power is proportional to $A = (a/\lambda)^4$, where A is the area and a is the radius of the aperture. In other words, the transmitted energy deceases in proportion to the fourth power of the electrical aperture size. A circular hole of diameter $1/100\ \lambda$ has an area that is $1/100$ the area of a hole of diameter λ. However, the small hole transmits $1/100 \times (1/10)^4 = 1$ millionth ($-120\,\text{dB}$) of the power that the larger hole transmits. Holes have very little effect if they are electrically small.

Mesh Shields and Faraday Cages

Due to the radiation limiting effects of small apertures, shields can be made from a mesh of wires or from a metallic cage, as an alternative to using a solid sheet of metal. From the results of apertures, you can deduce that the holes in the mesh or cage should be considerably less than half a wavelength for such shields to work properly. Moreover, the mesh will be mostly transparent to signals with wavelengths smaller than the mesh holes.

When the shield takes the form of a wire cage, it is often called a Faraday cage, named after its inventor, Michael Faraday. The Faraday cage principle is often used in the design of large satellite dishes as a means of saving money and reducing weight; the reflecting dish is constructed from a wire cage instead of a solid piece of metal. The spacing used for satellite dishes is often quite large, because the reflection does not need to be anywhere near 100%. The Faraday cage principle is also used where visual transparency is needed, as in the window of microwave ovens. Microwave ovens typically operate at $2.45\,\text{GHz}$ ($12\,\text{cm}$ wavelength). A typical microwave oven window shield has apertures of radius $a \cong 0.5\,\text{mm}$, corresponding to an electrical length of less than $1/100$. Assume that this shield is made of copper and has a thickness of about $20\,\text{mils}$. Further assume that there are $N = 50$ holes per cm^2. The theoretical transmitted power of this shield without the holes is about $-3000\,\text{dB}$. Without the holes, this shield is effectively ideal. You can now calculate the power transmitted in decibels from

$$T = 10\log_{10}\left[A_{holes} \times \left(\frac{a}{\lambda}\right)^4 \right]$$

SHIELDING IN THE FAR FIELD

where A_{holes} is the ratio of the hole area to the total area of the shield, $A_{holes} = N \times \pi a^2 \cong 0.4$. In other words about 40% of the shield is holes, and 60% is metal. Substituting the second formula into the first gives

$$T = 10\log_{10}\left[N \times \pi \times a^2 \times \left(\frac{a}{\lambda}\right)^4\right] \cong -100\,\text{dB}$$

As a decent approximation, in the far field (10 cm or so away from the screen), the microwave power is reduced by five orders of magnitude.

An important limiting factor for mesh screens is the thickness of the conductor portion in the mesh. Unless the frequencies are very high (i.e., skin depth is small), as in the case of microwaves, the conductor does not typically give the screen much depth. In such a case, the results given in the preceding paragraph, which assume a perfect shielding material, cannot be used. A further limitation is that inside a piece of electrical equipment, the circuitry is often placed fairly close to openings in the case. In other words, the far field approximation cannot be used. A commonly used approximation (Ott, 1988) for such applications is

$$T = 10\log_{10}\left[4 \times \left(\frac{\ell}{\lambda}\right)^2\right]$$

where ℓ is the largest dimension of the aperture.

A metal honeycomb structure can be used for applications that require robust shielding. The air intake for the fan of a product is an example where a solid sheet of metal cannot be used, but where very effective shielding may be necessary. A honeycomb or tubular shield can be used in any situation to decrease the power that passes through an aperture. For example, a single metal tube can be used to increase the shielding of a small aperture. These tubular shields are really waveguides operating below the cutoff frequency, a topic I will discuss later in the chapter. If the aperture is much smaller than a wavelength, the shielding effectiveness of a tubular aperture is approximately $S = 16z/a$, where a is the radius size of the openings and z is the depth of the tube.

Gaskets

Another method for reducing the leakage through shield openings is to fill the opening with metallic material. This technique works well for the seams where two pieces of an enclosure are fitted together. Several companies sell a variety of shielding gaskets designed for this purpose.

NEAR FIELD SHIELDING OF ELECTRIC FIELDS

In the near field, the wave impedance is no longer solely dependent on the intrinsic impedance of the air. Instead, the wave impedance is mostly dependent on the type of source and how far away the source is. For electric fields, the near field has a large electric component and a much smaller magnetic component. The corresponding wave impedance (ratio of electric to magnetic field) is very high. Since the wave impedance inside conductors is conversely very low, conductors act as excellent shields for electric near fields. Typically, sources of electric fields have relatively high voltage and low current. In other words, the sources themselves have high impedance. Dipole and monopole antennas, sparks, and high-impedance circuits produce near fields that are mostly electric. In the low-frequency limit (a DC electric field), electric field shielding becomes perfect. The electric field shielding effectiveness of copper shields is illustrated in Figures 9.6 and 9.7.

WHY YOU SHOULD ALWAYS GROUND A SHIELD

A shield does not need to be grounded to work. In fact, grounding a shield does not increase or decrease its shielding properties. So why then should shields be grounded? If the electronic device is completely enclosed, battery operated, and has no external cables, the shielding enclosure does not have to be connected to the circuit "ground" (i.e., the circuit common or neutral). However, if there are any cables that leave the box, an ungrounded shield can serve to capacitively couple signals from outside the box to inside the box. Furthermore, an ungrounded shield can act to couple signals between different circuits within the box even if there are no external cables. This problem doesn't diminish the shielding effectiveness but it does cause crosstalk within the product itself. Figure 9.8 demonstrates how this effect occurs. For these reasons, all shields should always be connected to the circuit neutral.

For safety reasons, any product powered by a wall outlet (i.e., mains voltage) should have its shielding enclosure connected to the grounding plug (the third prong) of the outlet. If an internal short circuit occurs whereby a high voltage comes into contact with the enclosure, the ground connection will prevent the enclosure from being brought to

Figure 9.6 Electric-field shielding: reflected, absorbed, and transmitted power for 10 mil thick copper shield.

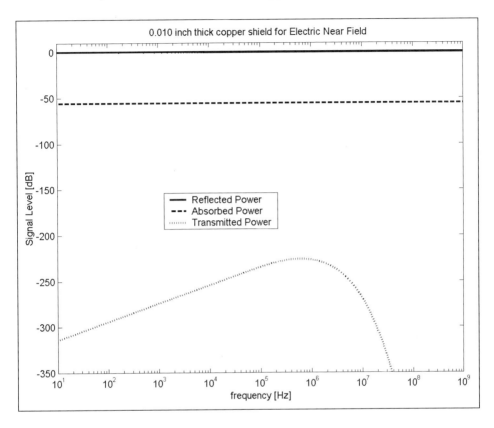

high potential. Without such protection, a high voltage could exist on the enclosure and cause a hazardous shock to anyone who happened to touch it.

NEAR FIELD SHIELDING OF MAGNETIC FIELDS

Magnetic fields are not reflected very well by conductors. A magnetic field has, by definition, a larger magnetic component than electric component. Therefore, its field impedance is very small. The intrinsic impedance of conductors is also small at low frequencies. Because the impedances are similar, not much reflection occurs. Absorption shielding must be used for magnetic fields. Consequently, to shield magnetic fields effectively, the shield must be several skin depths thick.

Figure 9.7 Electric-field shielding effectiveness for 10 mil thick copper shield.

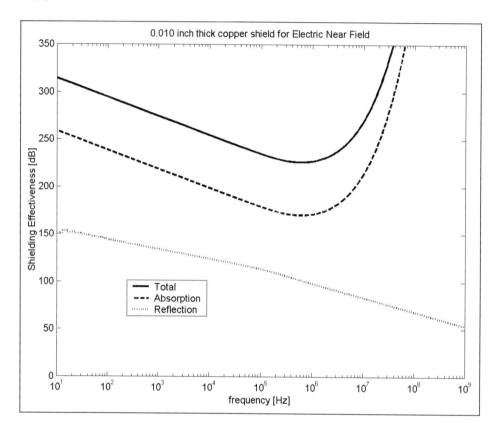

A shield's absorption effectiveness is dependent on the *electrical thickness* of the shield. The electrical thickness is most conveniently represented in terms of skin depths. Using the relationship, $f = v/\lambda$, skin depth can be expressed as

$$\delta = \sqrt{\frac{\lambda}{\pi v \sigma}}$$

Figure 9.8 Comparison of ungrounded to grounded shields. A circuit enclosed in an ungrounded shield (A) and the equivalent circuit (B). A circuit enclosed in a grounded shield (C) and the equivalent circuit (D).

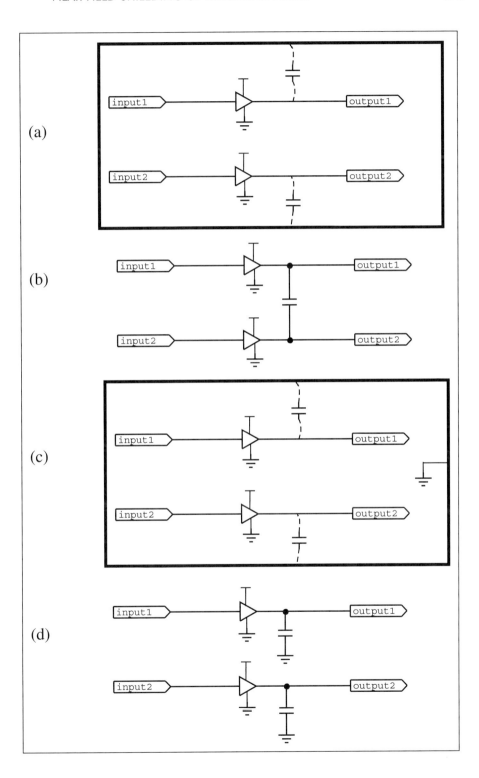

Hence electrical thickness is proportional to the square root of wavelength. A shield with an electrical thickness of 1 skin depth will allow $e^{-1} \sim 36.9\%$ of the incident radiation to be transmitted through the shield. In terms of decibels, the shield's effectiveness is $20\log_{10}[e^1]$ dB $\cong$ 8.7 dB; in other words, the transmitted signal is 8.7 dB less than the incident signal. The remaining radiation is absorbed. To create a good absorption shield, the shield should have an electrical thickness of several skin depths. The general equation for absorption loss is

$$A = -20\log_{10}[e^{-d_\lambda}] \cong 8.69 \times d_\lambda \text{ dB}$$

where d_λ is the electrical thickness of the shield (i.e., the number of skin depths).

Sources that have high current or low voltage produce near fields that are predominantly magnetic. In other words, sources of low impedance produce magnetic near fields. Motors and loop antennas are examples of circuits that produce magnetic near fields. As you can see from Figure 9.9 and Figure 9.10, magnetic shielding becomes completely ineffective as frequency goes to zero. At low frequencies, it is more practical to divert and concentrate magnetic fields by using shield materials with high permeability (such as iron or mu-metal) than it is to try to shield via absorption loss or reflection.

WAVEGUIDES

There exist three basic structures for transmitting a signal from one location to another: transmission lines, waveguides, and antennas. This chapter covers waveguides. A waveguide is a device that does just what its name implies—it guides radiated waves. Waveguides are typically only used at high frequencies (microwave and higher) because the cross-sectional diameter of a waveguide must be on the order of half a wavelength or greater for the waveguiding process to take place. In other words, the diameter must be electrically large for a waveguide to function. When the diameter is small, the radiation is cut off by diffraction effects similar to the effects of electrically small apertures. In microwave frequencies, waveguides take the form of hollow rectangular or cylindrical metal tubes. At light frequencies, waveguides take the form of solid cylindrical glass tubes—fiber optics.

Essentially, a waveguide forces a wave to follow its path by causing it to completely reflect at its boundaries. For example, suppose you construct a long metal tube with a rectangular cross-section of inside dimen-

Figure 9.9 Magnetic-field shielding: reflected, absorbed, and transmitted power for 10 mil thick copper shield.

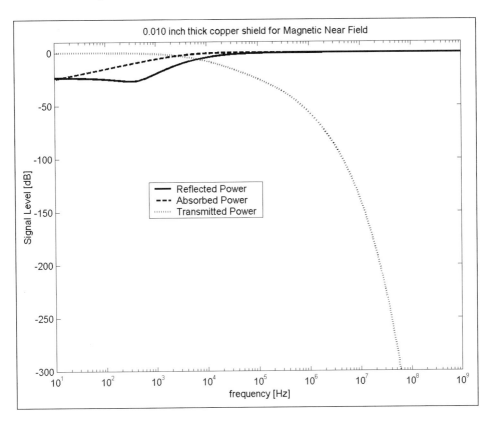

sions 2.84 inches by 1.34 inches (7.214 cm by 3.404 cm). These dimensions are the standard dimensions for an S-band microwave waveguide. Assuming that the plates are very well polished or are on the backside of a glass plate, you could even use this tube to guide the light from a laser or even a flashlight. Its operation is quite simple; if we think of the lights as rays, the rays are continually reflected by the walls of the tube, and they have net propagation only in the direction of the tube.

Cutoff Frequency

Since the metal tubes are good reflectors in microwave frequencies, you could place small antennas inside each end of the tube, and use it for a communication channel. For example, at 100 GHz, a quarter-wavelength monopole antenna has a wavelength of $\lambda = c/f \cong 3$ mm.

Figure 9.10 Magnetic-field shielding effectiveness for 10 mil thick copper shield.

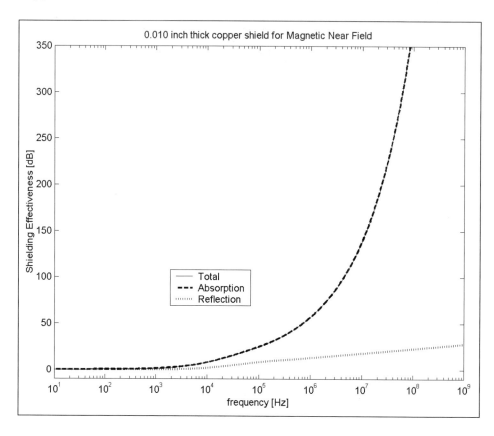

Therefore, it will be practical to place an efficiently radiating antenna inside this waveguide. You could use this tube for transmitting at even lower frequencies, but as you tune the frequency below about 2.078 GHz, you will observe your reception strength rapidly diminishing. At 1 GHz, you would effectively have lost the ability to transmit through the tube. The frequency below which transmission is cut off is aptly called the *cutoff frequency*—2.078 GHz in this example.

Multipath Transmission

There is an inherent problem with waveguide transmission, that of multipath transmission. If the waveguide aperture is much larger than a wavelength, the transmitted wave can take many different angled paths to get to the other end. Because each path has a different length, the waves of each will take a different amount of time to reach the receiving end. The outcome is that each path has a different effective velocity from transmitting end to receiving end. This type of velocity is called *group velocity*. The group velocity is the velocity at which the energy and information travel. If you try to send a signal from one end to the other, the power from the signal will split up and take many different paths. Because the paths have different velocities, some parts of the signal will arrive more quickly than others. The result is a signal with many echoes or ghosts. The term *ghosts* and *ghosting* are used when multipath occurs on a television set. You have probably seen this effect. The TV program has a faint, out-of-phase image superimposed on the normal image. This faint or ghost image is caused by some of the signal power reflecting off an obstacle and then being picked up by the antenna. The reflected signal travels a longer path and is delayed from the main image. This effect can also occur with cable TV when the cable transmission line is improperly matched, or if the cable has a defect in it somewhere. Reflections in the transmission line cause a ghost image to appear.

The process of multipath transmission is fairly easy to visualize for a parallel plate waveguide because it reduces to a reflecting ray in two dimensions. Tubular waveguides involve three-dimensional power flow paths, which are not easy to visualize.

Waveguide Modes

In the previous section I implied that the waves can follow an arbitrary reflecting path down the guide. This approximation can be used when the waveguide cross-section is much larger than the wavelength. However, in reality, there are only certain pathways that the waves can travel. The different pathways a signal can take through a waveguide are called *propagation modes* in electromagnetic terminology. Each mode has its own angle of propagation down the guide and hence its own group velocity. There are zero propagation modes below the cutoff frequency. If you choose a signal frequency that is just above cutoff, one propagation mode is available for transmission. If you raise the frequency a little higher, a second mode appears, and then a third mode, and so on. The modes form a series, and each mode has its own cutoff frequency below which it cannot propagate. Waveguides are operated at the frequency range just above the cutoff frequency of the lowest order mode. In this frequency range, there is exactly one propagation mode. One propagation mode correlates to one signal path and therefore no echoes or ghosts.

To explain why only certain modes are allowed, you need to learn some more about the physics of reflections from conductors. Assume that the conductor sheet is perfect or that it is several skin depths thick. Since a wave can only penetrate a small depth into the conductor, only a small amount of energy is transmitted to the conductor, and the rest of the energy must be reflected. Therefore, if we make the walls at least several skin depths thick, we get a good approximation of a perfect conductor (in terms of its ability to reflect waves). You may wonder what happens if the walls are too thin. In this case, the analysis used above does not apply, and the waveguide will not work properly. Much of the energy will leak out of the walls.

If a wave impinges upon the conductor sheet and is normally incident (at an angle of 90 degrees), reflection occurs that causes a standing wave. This standing wave is just like the standing wave that occurs on a short-circuited transmission line. The electric field goes to zero at the conductor surface and also in the air at every half wavelength from the conductor surface. If the wave arrives at an angle other than 90 degrees, the incident wave and reflected wave also create a standing wave. However, the placement of the nodes is now dependent upon the angle as well as the wavelength. The nodes occur at distance,

$$z = \frac{\lambda}{2\cos\theta} = \frac{c}{2f\cos\theta}$$

and at every multiple of this distance. As the angle is brought to zero, the nodes occur farther and farther apart.

Now consider two parallel conductor plates. In this case, both plates require the electric field to be zero at their surface. In other words, a standing wave node must occur at both plates. You have just learned that for a node to occur after a reflection, it must satisfy an equation that depends both on the angle of incidence and on the wavelength of the signal. If the signal frequency is just above the cutoff frequency, there exists only one angle for which this effect can occur, and the angle is very close to 90 degrees.

Figure 9.11 demonstrates graphically how this process occurs. As the frequency increases, other modes become available. For example, if you double the frequency to half the wavelength of the wave in Figure 9.11, there are now two angles at which the wave can propagate such that the boundary conditions are met. As the frequency is increased (wavelength reduced), more paths (modes) become available. Conversely, as frequency is reduced, fewer paths are available. When the wavelength becomes equal to the spacing of the plates, the wave must reflect at a

WAVEGUIDES

Figure 9.11 Graphical determination of standing wave nodes (dotted lines) by superimposing the incident and reflected waves.

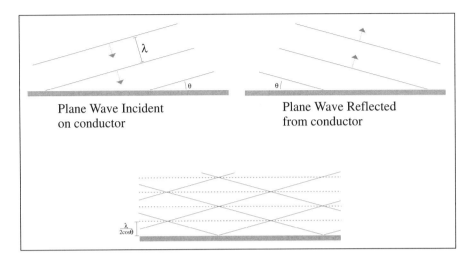

90 degree angle to the plates for it to meet the boundary conditions at the plates. At this angle, the wave just bounces up and down between the plates without ever propagating down the tube. This wavelength defines the cutoff wavelength (and cutoff frequency) of the guide. At this wavelength, the structure forms a resonant cavity, which I discuss later in this chapter. The fields of several waveguide modes are shown in Figures 9.12, 9.13, and 9.14.

Incidentally, the modes are referred to by labels like TE_1 or TM_2. In practice, unless the cross-section is very large compared to the wavelength, two waves are excited for each mode. At any point in space, one wave is traveling upward and one wave is traveling downward. The upward and downward waves of each mode sum in such a way that the resulting wave always has either the electric or magnetic field pointing in a direction that is transverse (perpendicular) to the tube walls. The TE/TM portion of the label refers to which of the fields (electric or magnetic) is transverse to the direction of propagation down the guide. Every waveguide has a series of TE modes and a series of TM modes. The numbers are integers that refer to the mode number. For example, TE_1 refers to the first transverse-electric mode.

Waveguide Modes on Transmission Lines

For the most part, I have talked about true waveguides in this chapter (i.e., tubes for guiding radiating waves). Except for the parallel plate

Figure 9.12 Waveguide operating just above cutoff frequency ($f = 1.0353\ f_{cutoff}$, $\theta = 15°$).

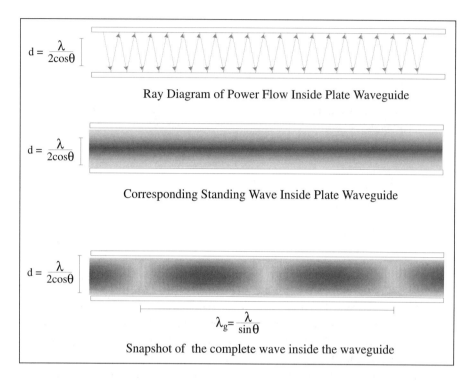

guide, these waveguides are characterized by the fact that there is only a single conductor—the walls of the tube. In fact, any of the two-conductor transmission lines that have flat, plate-like, conductors can also carry waveguide modes at high frequencies. In waveguide terminology, the transmission line mode or circuit mode of these structures is called the TEM (transverse electric and magnetic) mode. The TEM mode is the only mode that can propagate at DC and at low frequencies. The commonly used transmission line types that support waveguide modes are coaxial cables, microstrip, and strip-line. To avoid multimode phenomenon, these transmission lines should only be used below the frequency of the first waveguide mode. The first waveguide mode appears between 10 GHz and 100 GHz for most coaxial cables. A coaxial cable should not be used at or above this frequency. The microstrip and strip-line transmission lines used on printed circuit boards also have frequency limitations due to the onset of waveguide modes. Microwave engineers and very high-speed digital designers (>10 GHz) need to be aware of this

Figure 9.13 Waveguide operating in the first mode (TE$_1$ or TM$_1$) at twice the frequency of that in Figure 9.12 (f = 2.0706 f_{cutoff}, θ = 61.1°).

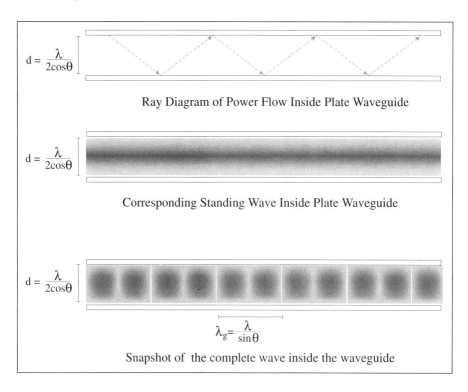

problem and use thin dielectric spacing between the trace and ground plane that comprise the transmission line.

The Earth's Waveguide

The Earth itself has its own waveguide—that formed by the ground and the ionosphere. The ionosphere is a plasma (i.e., ionized gas) that acts to reflect low-frequency radio waves such as those used in AM-band radio, and for many amateur radio bands. The ground also acts as a decent conductor for low-frequency radio waves. The ionosphere and the ground thus form a curved "parallel plate" waveguide. Sometimes you can receive radio broadcasts from extremely long distances because the signal has reflected from the ionosphere and back down to the Earth's surface to your receiving antenna. This type of radio transmission is called *skip-mode* transmission. However, skip-mode transmission is not very reliable because the ionosphere changes with weather,

Figure 9.14 Waveguide operating in the second mode (TE$_2$ or TM$_2$) at twice the frequency of that in Figure 9.12 (f = 2.0706 f_{cutoff}, θ = 15°).

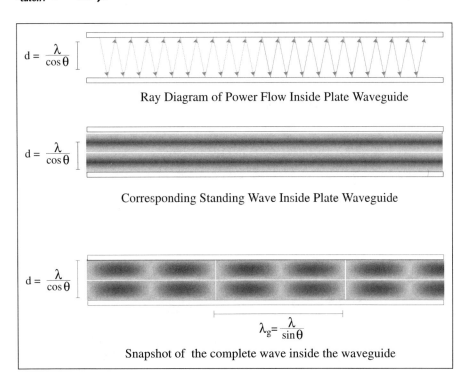

season, and time of day. Typically, at night, frequencies under 2 MHz can propagate via skip mode. During the daytime, frequencies under 10 MHz can typically propagate via skip mode. For frequencies under 150 kHz, the ionosphere is almost always an excellent reflector of waves, and is mostly independent of atmospheric effects. Signals in this band can propagate vast distances (thousands of miles) and are thus used for navigation and time signal transmissions. For radio waves that are less than 2 MHz, the energy can also travel past line of sight via a ground surface wave, as discussed in the next chapter.

Antennas for Waveguides

There are two common methods for exciting waves in hollow metal waveguides. Both forms involve using the center conductor of a coaxial cable as an antenna. Figure 9.15 shows the two basic antenna feeds

Figure 9.15 Waveguide feed examples using coaxial cable. A) This feed system excites a wave via an electric monopole antenna. **B)** This feed system excites a wave via a magnetic loop antenna. In both figures, the outer conductor (shield) of the coax is connected to the waveguide wall.

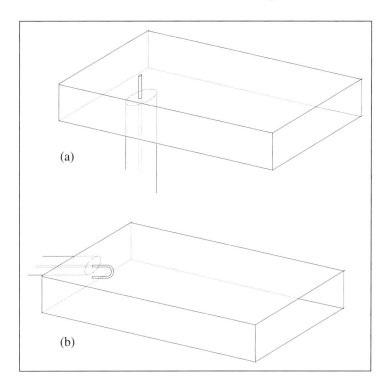

for waveguides. In the first example, the center conductor projects out past the outer ground and forms a monopole antenna. (Note that the center conductor does not touch the top of the guide.) In the second example, the center conductor extends out past the ground and then loops back to connect to the ground, forming a simple loop antenna.

Evanescent Waves and Tunneling

When the cross-section of a waveguide has dimensions smaller than half of a wavelength, waves cannot propagate. But the fields do penetrate a small distance into the guide, diminishing in an exponential manner. Such fields are called *evanescent waves* and are similar to the reactive storage fields that occur around circuits. In optics, such evanescent waves are referred to as *frustrated waves*. These waves have the interesting property that they can propagate energy if they are short and if they couple to larger regions. For example, consider a waveguide operating at the fundamental mode. Now suppose you compress the metal waveguide within a small region such that the waveguide has a neck in it that is smaller in cross-section than the cutoff wavelength of the operating signal. If this region is short enough, some amount of the energy can actually couple through the evanescent region. This effect is common to all wave phenomena. In quantum mechanics, such behavior is referred to as *tunneling*. Quantum tunneling is often given a mystical status, as if it only occurs in the quantum world. In fact, tunneling is a universal wave phenomenon and in quantum physics is just a consequence of the wave-particle duality of matter.

RESONANT CAVITIES AND SCHUMANN RESONANCE

Resonant cavities are similar to waveguides. Just as solid cavities, such as organ pipes or glass bottles, support sound resonance, metal cavities support electromagnetic resonance. The resonant frequency of a metal cavity is simply the frequency whose wavelength corresponds to the dimensions of the box or cavity. A wave is set up inside the box whose nodes (points of zero amplitude) lie on the walls of the box. Resonant cavities are used in the microwave band for creating resonators for oscillators and filters. I have even heard stories of amateur radio operators using metal garbage cans as resonant cavities in the high VHF band!

The Earth itself forms a resonant cavity. The ground and the ionosphere of the entire planet form a giant spherical cavity whose resonance is at about 10 Hz. This resonance is called *Schumann resonance* after its discoverer. This resonant cavity can be excited by lightning strikes. A lightning strike is basically a giant spark. As such, it is a transient phenomenon, producing a pulse containing many frequencies. You could in theory use this resonance to track lightning strikes; however, actual lightning tracking systems currently in use utilize the higher, waveguide frequencies of the Earth in the range of 1 kHz to 1 MHz.

FIBER OPTICS

Another method for guiding waves is the use of solid dielectric tubes. Typically such guides are made from glass fibers and/or plastics. The fibers usually are constructed from two slightly different materials. The inner portion of the fiber, called the core, consists of a material of high dielectric constant. In optics, we typically refer to the index of refraction instead of the dielectric constant. The two parameters describe the same material property. The index of refraction, n, is simply the square root of the relative dielectric constant,

$$n = \sqrt{\frac{\varepsilon}{\varepsilon_o}}$$

The outer portion of the fiber, called the cladding, consists of a material of slightly lower dielectric constant. The boundary of the two materials is an impedance boundary, where reflections take place. If the angle is slight, all of the incident light is reflected. This phenomenon is quite familiar to most people. Looking straight ahead at water or glass, you see little reflection. However, at an angle, the reflected light is great. Glass rods or fibers can also be used for guiding radio waves, but at these frequencies, metal guides are more practical and are used instead. The application of dielectric rods for guiding radio waves was actually studied as early as 1910, long before the laser was invented.

The problem of multimodes occurs in fiber optic guides in addition to metal guides. The multimode problem is overcome in fiber optic communication because information is sent in on/off pulses of laser light. The effect of multimodes causes the pulse to broaden or disperse, but communication is still possible, as long as enough time is provided between pulses.

LASERS AND LAMPS

The laser provides a very narrow bandwidth of waves and can be turned on and off very quickly (at GHz to THz rates) allowing extremely high data rates. Multimode fibers also allow for lasers of several different wavelengths (slightly different color) to transmit and receive on the same fiber. Such techniques are called wavelength division multiplexing (WDM). At present, hundreds of wavelengths can be transmitted on the same fiber.

The laser can be thought of as an optical oscillator. Whereas radio equipment and digital systems use transistors to produce a single frequency, the laser uses quantum transitions of electrons in semiconductors, which occur at exact frequencies. Such light is referred to as coherent light (single frequency and in phase) in optical literature.

In contrast to lasers, incandescent lamps are "noise." Their light is produced by random thermal agitation of the electrons in hot materials. As such, the light of lamps is broadband in frequency and random in phase. Animal vision is based on average intensity of the light. Our eyes utilize only the amplitude of the light, and not the phase. The mix of different wavelengths has the effect of producing the different colors of our visual perception.

BIBLIOGRAPHY

Shielding

Bethe, H. A., "Theory of Diffraction by Small Holes," *The Physical Review*, Vol. 66, Nos. 7 and 8, October 1 and 15, 1944.

Morrison, R., *Grounding and Shielding Techniques*, 4th Edition, New York: John Wiley, 1998.

Ott, H. W., *Noise Reduction Techniques in Electronic Systems*, 2nd Edition, New York: John Wiley, 1988.

Paul, C. R., *Introduction to Electromagnetic Compatibility*, New York: John Wiley & Sons, 1992.

Williams, T., *EMC for Product Designers*, Oxford: Butterworth–Heinemann Ltd, 1992.

Waveguides

Carr, J. J., *Practical Antenna Handbook*, Fourth Edition, New York: McGraw-Hill, 2001.

Collin, R. E., *Field Theory of Guided Waves*, 2nd edition, IEEE Press, 1991.

Heald, M., and J. Marion, *Classical Electromagnetic Radiation*, 3rd Edition, Fort Worth, Tex.: Saunders College Publishing, 1980.

Jackson, J. D., *Classical Electrodynamics*, 2nd Edition, New York: John Wiley & Sons, 1975.

Johnson, H., "Strange Microstrip Modes," *EDN*, April 26, 2001.

Jordan, E. C., and K. G. Balmain, *Electromagnetic Waves and Radiating Systems*, 2nd Edition, Englewood Cliffs, N.J.: Prentice Hall, 1968.

Karbowiak, A. E., "Theory of Imperfect Waveguides: The Effect of Wall Impedance," *Radio Section*, No. 1841, London: Imperial College, September 1955.

Karbowiak, A. E., "Some Comments on the Classification of Waveguide Modes," *The Proceedings of the IEE*, Vol. 107, Part B, No. 32, March 1960.

Kraus, J. D., and D. A. Fleisch, *Electromagnetics with Applications*, 5th Edition, Boston: McGraw-Hill, 1999.

Paul, C., and S. Nasar, *Introduction to Electromagnetic Fields*, 2nd Edition, Boston: McGraw-Hill, 1987.

Pozar, D. M., *Microwave Engineering*, 2nd Edition, New York: John Wiley, 1998.

Ramo, S., J. R. Whinnery, T. Van Duzer, *Fields and Waves in Communication Electronics*, 2nd Edition, New York: John Wiley, 1989.

Shrader, R. L., *Electronic Communication*, 5th Edition, New York: McGraw-Hill, 1985.

Wave Propagation on Earth and in the Atmosphere

Boithias, L., *Radio Wave Propagation*, Boston: McGraw-Hill, 1987.

Carr, J. J., *Practical Antenna Handbook*, Fourth Edition, New York: McGraw-Hill, 2001.

Cummins, K. L., E. P. Krider, M. D. Malone, "The U.S. National Lightning Detection Network and Applications of Cloud-to-Ground Lightning Data by Electric Power Utilities," *IEEE Transactions of Electromagnetic Compatibility*, Vol. 40, No. 4, November 1998.

Jackson, J. D., *Classical Electrodynamics*, 2nd Edition, New York: John Wiley & Sons, 1975.

Jordan, E. C., and K. G. Balmain, *Electromagnetic Waves and Radiating Systems*, 2nd Edition, Englewood Cliffs, NJ: Prentice Hall, 1968.

Kraus, J. D., and D. A. Fleisch, *Electromagnetics with Applications*, 5th Edition, Boston: McGraw-Hill, 1999.

Orr, W., *Radio Handbook*, 32nd Edition, Woburn, Mass.: Butterworth–Heinemann, 1997.

Straw, R. D., Editor, *The ARRL Antenna Book*, 19th Edition, Newington, Conn.: American Radio Relay League, 2000.

Web Resource

http://ewhdbks.mugu.navy.mil/WAVEGUID.htm

10 CIRCUITS AS GUIDES FOR WAVES AND S-PARAMETERS

In the previous chapter, I mentioned that waves do not propagate well in conductors. They attenuate down to miniscule levels after traveling a few skin depths. Keep in mind that the conductor actually becomes more lossy at higher frequencies because the current must travel through a smaller cross-sectional area. High-voltage power lines often use multiple parallel wires for each phase partly because of the skin effect. At 60 Hz, the skin depth is about 8 mm. By splitting the current into several wires, you get more surface area for an equal amount of copper. Such an arrangement, called conductor bundling, minimizes costs for producing a low-resistance cable. Having several wires for each phase also helps reduce corona effects because the fields are not as concentrated. The reduction of corona losses is the main reason for the use of conductor bundling. The technique of conductor bundling is also used for winding RF transformers. Special wire, called Litz wire, is used to reduce problems associated with the skin effect (higher resistance and inductance). Litz wire consists of several enameled conductors bound and twisted together. By coating each wire with a dielectric, the effective surface area is increased.

Electromagnetic waves also travel extremely slowly in conductors:

$$v = 2\pi f \times \delta$$

where δ is the skin depth in meters. In air, electromagnetic waves, whether 60 Hz power line waves, light waves, or radio waves, travel at about the speed of light in a vacuum, 3×10^8 m/sec or 670 million miles/hour. In copper at 20 Hz, electromagnetic waves travel at about 2 m/sec or 4 miles per hour! To be clear, I am talking about the waves, not the electrons. (As mentioned in Chapter 2, the electrons travel even slower, with average drift velocity of about 1.7 miles/hour in copper.)

Take the example of a telephone line. Assume that the phone company is three miles from your house, and the phone company sends the signal to ring your phone. (The ring signal is a 20 Hz sine wave in North America.) This signal will take about forty-five minutes to travel the three miles through copper. In addition, the wave will be attenuated to immeasurable levels. Experience tells you that something must be wrong with the calculations, but I assure you they are quite correct. What is wrong is the assumption that the electromagnetic wave travels from source to destination on the inside of the wire. In reality, the wave travels at the surface of the wires and in between the wires. There is, of course, a wave inside the conductor, but this wave actually propagates inward from the surface of the wire. Moreover, the wave travels very slowly into the wire and attenuates due to the skin effect. The wave will propagate a considerable distance down the wire, perhaps even the entire 3 miles to the phone, in the same time that it takes for the wave to reach the center of the wire.

From measurements, we know that waves travel at about the speed of light on transmission lines, and that the wavelength along transmission lines is on the same order as that in air—further proof that signals are not carried on the inside of the wires, but at the wire surfaces and in between the wires. The actual function of the wires is to guide the wave energy between the wires. The wave travels along the surface and penetrates inward. The myth of electricity traveling through the wires like water through a pipe is really just that, a myth. Of course this myth is very useful for electronic circuits. Trying to design and analyze circuits with the techniques of electromagnetics would be very cumbersome, but understanding what actually happens in the circuits at high frequencies is very important.

SURFACE WAVES

Surface waves are another phenomenon common to all wave types. Surface waves occur at the boundary between two materials and travel along the surface. Moreover, the energy of surface waves is concentrated in the space close to the boundary. Water waves, such as ocean waves, are a familiar example of surface waves. Acoustic waves in solids can also form surface waves. The propagation of earthquake waves along the surface of the earth takes the form of a surface wave called a head wave. Surface acoustic wave (SAW) devices are very important in many wireless applications because they are compact and can operate into the GHz frequency range. Such devices utilize acoustic waves that travel along the surface of a piezoelectric material, typically quartz, and can be excited via electronic signals applied to specially designed electrodes.

Figure 10.1 1 MHz surface wave on copper, traveling to the right (*top*). Closeup of wave at copper surface (*bottom*).

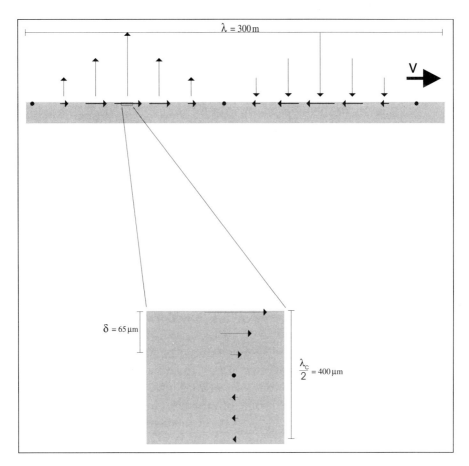

Many types of electromagnetic surface waves exist. One type of surface wave occurs when an antenna is placed over a conducting surface such as copper. Such a wave has an orientation as shown in Figure 10.1. The wave is propagated via the lateral movement of charge at the surface. Inside the conductor, the electrical field is mainly horizontal and the current flows along these lines. Outside the conductor, the electric field is mainly vertical, caused by the charged regions in the wave.

There is no net transportation of charge in this process. After each cycle, the charge ends up at the place where it started. A salient feature of the surface wave is that the majority of the current is within one skin depth of the surface. The currents in the conductor dissipate wave energy as heat, in accordance with Ohm's law. Although there is no net movement of any charge, at any given moment there is displacement of charge. Hence there are regions of positive, negative, and neutral charge. It is these regions of net charge that cause the vertical field lines outside the conductor.

The magnetic field points in the same direction in both the air and the conductor. The magnetic field is parallel to the boundary and points directly into or directly out of the page, depending on whether the current is flowing to the left or to the right, respectively. This wave is propagating to the right at a speed very close to the speed of light. The conductor slows it down a small amount and the speed varies slightly with frequency,

$$v \approx \frac{c}{1+\dfrac{1}{8}\left[\dfrac{2\pi f \times \varepsilon_{conucutor}}{\sigma_{conductor}}\right]^2}$$

where the approximation is valid for good conductors. For copper, the wave travels at essentially the speed of light. Even at 100 GHz, the speed is 99.99999% of the speed of light. As it propagates, the wave carries energy with it to the right. At the same time, it also propagates energy down into the conductor where it is converted to heat by the flowing currents.

This type of surface wave is produced by radio antennas and propagates along the surface of the Earth, as shown in Figure 10.2. Such a wave is often called a Zenneck wave or a Norton wave, named after two of the men who produced some of the initial theoretical work. In 1907, Zenneck proposed that such a wave explained the discovery of radio waves that could travel tremendous distances around the globe. It was later found that extremely long-range radio communication was due to radio waves reflected by our planet's natural ionosphere/ground waveguide. Nonetheless, such surface waves do indeed occur along the Earth's surface. Norton was one of the researchers who developed the full, correct equations for the surface wave produced by an antenna above the Earth.

The Beverage or Wave antenna, invented in 1923 by an amateur radio operator, is an antenna that is specially adapted to transmit and receive surface waves along the Earth. This type of antenna consists of a horizontal wire close to the ground. Such an antenna must be several wavelengths long to operate effectively, and placed close to the ground (less than 0.1λ). The wave is transmitted (or received) in the same direction as the wire, as opposed to transmitting broadside to the antenna like a half-wavelength dipole. The Wave antenna can function efficiently up

Figure 10.2 1 MHz surface wave on soil, traveling to the right. The arrows in the air represent about 10 times the magnitude of the arrows in the soil.

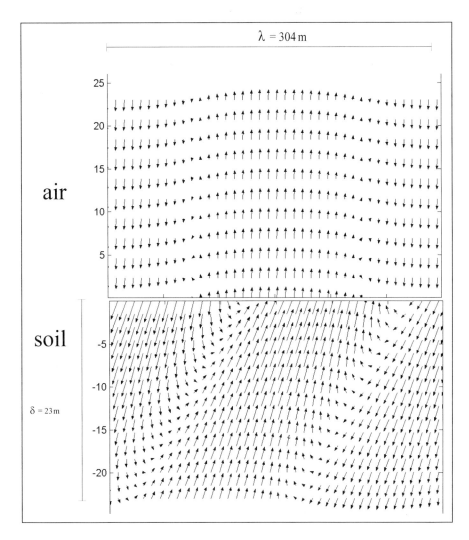

to about 2 MHz. There has been much controversy concerning the operation of the Wave antenna. A proper treatment is given in King (1983).

SURFACE WAVES ON WIRES

A similar type of surface wave can be excited on a single cylindrical wire. Contrary to popular belief, a single wire can serve to transport electrical signals! Let me be very clear about this; I am not referring to "ground return" circuits that use the ground as a second conductor for signal transmission as in early telegraph and telephone systems. I am referring to a Sommerfeld-Goubau wave, which is guided by a single wire without coupling to ground or any other conductors. This type of wave can be excited by extending a single wire from the center conductor of a coaxial cable. At the end of the coaxial cable, the shield can be flared into a cone to facilitate launching the wave onto the single wire. A reciprocal setup can be used at the receiving end. The wave has the same characteristics as the surface wave on a plane, except that it encircles the wire. It travels along the surface of the wire and propagates energy along the wire, as well as propagating some energy into the wire. Here again, the energy that propagates into the wire is dissipated as heat, and its penetration into the wire is thus limited by the skin effect. This type of wave was heavily investigated during the 1950s in hope of creating long distance transmission lines with half the amount of copper. The idea failed to be practical, however, because the signals were dramatically affected by the environment surrounding the wire. Birds sitting on the wire disrupted signals, and rain and dirt caused terrible amounts of attenuation. The wire must also be extremely straight, or the signal is radiated away. Nonetheless, the fact that such a wave can be produced is still amazing to ponder.

COUPLED SURFACE WAVES AND TRANSMISSION LINES

Traditional transmission lines in practice consist of two or more conductors. In a two-wire transmission line, one wire is typically designated the signal wire, and the other is designated the return or ground wire. Surface waves also provide the transport of the signal and its associated energy for these structures and for all electronic circuits in general. This wave consists of two cylindrical surface waves, one in each conductor, which are closely coupled to one another. For proper operation (i.e., single mode, low radiation) of a transmission line, the wires must be placed closed together (much less than a wavelength in spacing), so that the surface wave can couple via the near field. The surface waves also serve to create regions of net positive and negative charge, which creates a wave between the conductors. The wave between the conductors is usually called a TEM (transverse electric and magnetic) wave because its components are both approximately perpendicular to the direction of propagation. Barlow and Cullen (1953) is a good reference for exploring the detailed derivation of the relation between surface waves and the coupled surface waves of transmission lines.

COUPLED SURFACE WAVES AND TRANSMISSION LINES

Like other surface wave structures, transmission lines carry the power in the air (or dielectric) between the wires. As the wave travels along the wires, it also propagates power (at a very slow speed) into the wires where it dissipates via the wire resistance. The depth of penetration into wires is determined by the skin depth, as described in Chapter 9. The velocity of the wave along the wire is affected by the resistance of the wires and by the leakage conductance between the wires. It can be calculated from the per length parameters of the transmission line. If the losses are small, the velocity can be approximated by

$$v = \frac{c'}{1 - \frac{1}{4}\left[\frac{RG}{2\pi f \times LC}\right]^2 + \frac{1}{8}\left[\frac{G}{2\pi f \times C}\right]^2 + \frac{1}{8}\left[\frac{R}{2\pi f \times L}\right]^2}$$

where R is the wire resistance, G is the dielectric conductance, L is the line inductance, C is the line capacitance, and c' is the speed of light in the dielectric. Notice the similarity between this equation and the surface wave equation for velocity.

How a Transmission Line Really Works

As an example of how the surface waves propagate energy along the transmission line, refer back to the example of a battery connected to a transmission line via a switch, as given in Chapter 8. Before the switch is closed, positive charge will have built up on the left side of the switch. On the right side of the switch, negative charge will have built up in exact opposition to the positive charge. To keep things simple, assume that the mobile charge carriers are positive charges. When the switch is closed, the positive charges to the left of the switch propagate across the connection abruptly. These charges undergo a pulse of acceleration, producing a field kink (Figure 5.5C) that propagates away from the charges at the speed of light. Being in the near field, the positive charges on the lower conductor experience this pulse as a very strong force, as shown in Figure 10.3. Notice that the field in the pulse region is directed to the left. Thus, positive charges in the bottom wire will now be abruptly accelerated to the left. These charge send out their own field kink, which pushes the charges on the top conductor to the right. In effect this is a reinforcing feedback system; consequently, the charge movement in the two wires quickly becomes strongly coupled. The pulse wave progresses away from the switch and down the transmission line, carrying a wave of energy toward the load. Keep in mind that since waves propagating through the bulk of the wire travel very slow and experience rapid attenuation, the charge movement described

Figure 10.3 A) A pulse traveling down a transmission line.* The direction of current is shown by the gray arrows. The top charge moves to the right; the bottom charge moves to the left. The black arrows show the electric field caused by each moving charge. As you can see, the fields of each charge reinforce the movement. (The charges are assumed to be positive [+] for simplicity.) The charges carry this pulse of energy to the right. B) The electric field lines after steady state has been established. Inside the wire, the electric field lines point in the direction of the current. The wave impedance between the wires of the transmission line is equal to the characteristic impedance of the line. The wave impedance in the neighborhood of the load resistor is equal to the resistance of the load. To see an "end-on" view of the field surrounding the load, refer to Figure 2.15.

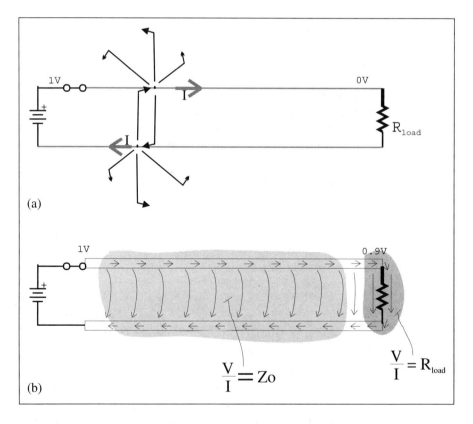

*This is an exaggerated diagram, as it shows the electric field from a far-field perspective. The coupling actually takes place via a near-field coupling, and the "kinking" of the field has a much more subtle appearance. However, the conceptual framework of this figure is correct.

occurs at the surface of the wires. Furthermore, the wave energy is carried in the field kink between the two wires. The initial wave can actually reach the load before it reaches the center of the conductor.

As the wave travels along, the ratio of the voltage to current is equal to the characteristic impedance of the line. When the wave reaches the end of the transmission line, the wave impedance changes because here the ratio of the voltage to current is determined by the impedance of the load. In the same way that freely propagating waves, such as light, reflect when an impedance boundary is encountered, so too is this coupled surface wave reflected.

Transmission Line Modes and Waveguide Modes

Although transmission line modes (the waves of ordinary electronic circuits) and waveguide modes are both fundamentally guided waves, there is an important difference between the two types of waves. The transmission line modes can propagate energy at any frequency, including DC. To propagate energy down to DC, transmission line modes can occur only on structures with two or more conductors. Transmission line modes are also characterized by near field (nonradiating) coupling of energy. For a transmission line to operate, the conductors must be placed electrically close together, so that near field coupling can take place. If this rule is not followed, the transmission line becomes an antenna.

On the other hand, waveguides can only propagate energy above their cutoff frequency, which is determined by the waveguide geometry and cross-sectional dimensions. The cross-sectional dimensions must be large enough to allow a radiating wave to propagate inside the guide. The waveguide works by forcing a radiated wave to travel through the guide by reflecting the wave at the waveguide boundaries. Because the waveguide works via reflection, it can take many forms: closed single-conductor metal tubes, dielectric fibers, two-conductor systems with flat surfaces such as coaxial cable, and microstrip, as well as other types. *Simply stated, transmission lines propagate waves via coupling of near field energy, and waveguides propagate waves via reflections of radiated energy.*

LUMPED ELEMENT CIRCUITS VERSUS DISTRIBUTED CIRCUITS

At high frequencies parasitics start to dominate the behavior of lumped circuit elements such as resistors, transistors, capacitors, and inductors. As the wavelength of the signal becomes small enough to be comparable in size to the devices, lumped element design becomes impossible. On printed circuit boards, state of the art, surface mount device (SMD) technology allows for designs to reach frequencies of several GHz. As components become even smaller, this frequency limit will increase. Inside integrated circuits (ICs), devices can be made that are several orders of magnitude smaller than the smallest discrete component. Devices with dimensions of less than 1 micron (10^{-6} meters) can now be manufactured. These tiny devices have pushed the frontier of electronics into the terahertz range.

An alternative to miniaturization techniques is the use of distributed element design techniques. Such techniques are commonly used at microwave frequencies, where lumped discrete elements are not available. The microwave frequency range is formally defined as 300 MHz to 300 GHz, which corresponds to wavelengths of 1 m to 1 mm. In modern practice, the frequency range from 300 MHz to 1 GHz is usually referred to as UHF, with the term *microwaves* being reserved for frequencies above 1 GHz. Frequencies above 100 GHz are often referred to as millimeter waves, and frequencies above 300 GHz are referred to as submillimeter waves or far-infrared. With the advent of surface mount devices, distributed microwave techniques are typically not necessary unless you are designing above 1 GHz.

Figure 10.4 shows examples of distributed design techniques. Each of these examples is based on the use of microstrip transmission lines. A microstrip transmission line is the commonly used transmission line on printed circuit boards for both RF and digital applications. It consists of a copper trace placed over a ground plane. The first example shows the creation of a series capacitance by placing a small gap in a microstrip trace. The second example shows inductive coupling (i.e., a transformer) by placing two microstrip transmission lines in close proximity. The third example shows a band pass filter constructed from transmission lines. The center trace is exactly a quarter wavelength at the resonant frequency. The input and output are inductively and capacitively coupled to the resonant piece in the middle. Like most distributed resonators, the center trace is also resonant at every harmonic of the fundamental frequency.

$\lambda/8$ TRANSMISSION LINES

Another distributed technique is the use of $\lambda/8$ length transmission lines in place of lumped elements. In Chapter 7, you learned that $\lambda/4$ transmission lines had special properties that transform a short to an open and vice versa. Transmission lines of length $\lambda/8$ also have special properties.

Figure 10.4 Distributed circuit techniques with microstrip transmission lines. Each figure shows the top view of the physical layout. A continuous ground plane is placed below each circuit.

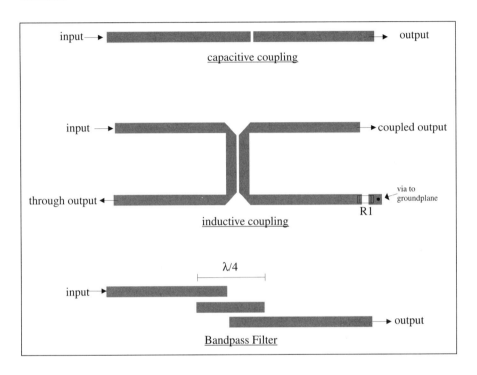

A short-circuited λ/8 transmission line behaves as an inductor of value $L = Z_o/2\pi f$. Similarly, an open-circuited λ/8 transmission line behaves as a capacitor of value $C = 1/(2\pi f \times Z_o)$. This technique can be used for creating parallel capacitors and inductors by connecting a λ/8 transmission line stub to the main transmission line (Figure 10.5). In general, the characteristic impedance of the stub is not the same as the characteristic impedance of the main line. Instead, the stub impedance is set to produce the proper capacitance or inductance. Notice that both equations are dependent on the frequency of operation. Being a distributed technique, λ/8 transmission lines are a narrow band technique, which can only be used in the frequency range near the frequency $f_0 = c/\lambda$.

S-PARAMETERS: A TECHNIQUE FOR ALL FREQUENCIES

When you start working with distributed circuits, transmission lines, waveguides, antennas, and other high-frequency elements, the tradi-

Figure 10.5 Distributed circuit elements using λ/8 stubs. A) Microstrip capacitor using an open stub, and its equivalent circuit. The stub width determines the capacitance. B) Microstrip inductor using a shorted stub, and its equivalent circuit. The stub width determines the inductance.

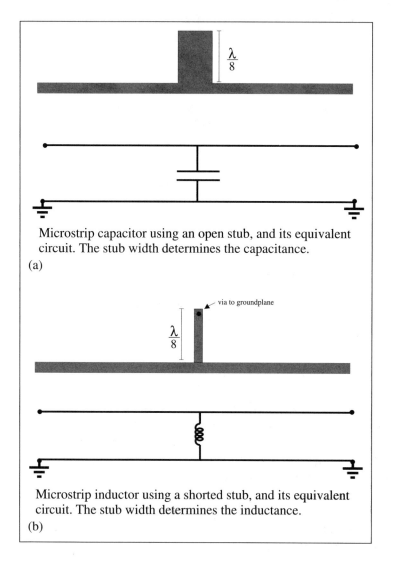

Microstrip capacitor using an open stub, and its equivalent circuit. The stub width determines the capacitance.
(a)

Microstrip inductor using a shorted stub, and its equivalent circuit. The stub width determines the inductance.
(b)

tional circuit techniques of Kirchhoff's voltage and current laws either lose their utility (e.g., transmission lines) or have no applicability (e.g., waveguides). One method that can be universally used is the direct application of Maxwell's equations. This technique is indeed used in electromagnetic circuit simulators, some of which can actually extract circuit parameters from 3D models. However, obviously the use of Maxwell's equations is much too cumbersome for most analysis.

To analyze electromagnetic devices at virtually any frequency, the technique of "scattering" parameters, commonly called *S-parameters*, was invented. The technique of S-parameters uses the concept of wave propagation for describing devices. The S-parameters relate to the reflection and transmission coefficients of the device, relative to a given characteristic impedance. Since electronic devices are typically frequency dependent, S-parameters for a device will be, in general, a function of frequency. S-parameters are also based on the concept of ports. Each device is characterized by its number of ports. Resistors, capacitors, antennas, and voltage sources are examples of *one-ports*. Transmission lines, waveguides, amplifiers, and filters are examples of *two-ports*. For circuits, each port corresponds to a wire pair leaving the device. For devices in general, each port corresponds to an input/output interface to the device.

One-port devices can be classified into two types: loads and sources. A one-port load has one parameter, S_1, which is just equal to the voltage reflection coefficient,

$$S_1 = \Gamma = \frac{Z - Z_o}{Z + Z_o}$$

where Z is the impedance of the device. A one-port source is defined by two parameters: the reflection coefficient, S_1, and the amplitude of the source voltage wave, V_s. For an AC circuit voltage source of voltage V_{supply}, the source voltage wave is the value of the output voltage when the source is terminated with a load impedance equal to the source impedance. The resulting relation is simply

$$V_s = \frac{V_{supply}}{2}$$

The S-parameters for a two-port device form a 2 × 2 matrix,

$$S = \begin{bmatrix} S_{11} & S_{12} \\ S_{21} & S_{22} \end{bmatrix}$$

The coefficients S_{11} and S_{22} are the reflection coefficients at port 1 and port 2, respectively. The coefficient S_{12} is the transmission coefficient from port 2 to port 1 (backward gain). The coefficient S_{21} is the transmission coefficient from port 1 to port 2 (forward gain). For an

Figure 10.6 The directional coupler and examples of its use. Don't take the port names literally; the names are meaningful only when a signal is applied to port 1.

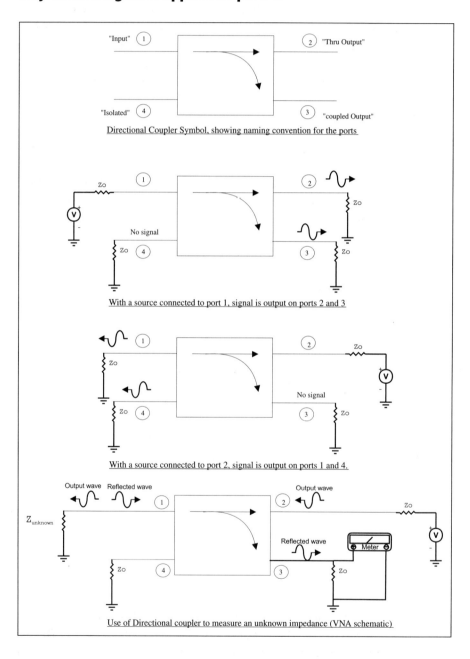

Figure 10.7 Implementation of a microstrip directional coupler, specifically a 90 degree quadrature hybrid. The input is split equally between the two outputs. The lower output is 90 degrees out of phase with the top output.

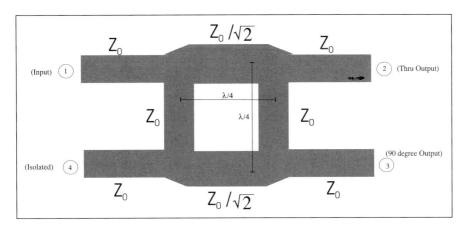

amplifier, S_{11} and S_{22} correspond to the input and output impedance, respectively. Parameter S_{21} specifies the gain of the amplifier, and parameter S_{12} is zero. You need to be careful of the tendency to oversimplify the use of two-port S-parameters. For example, parameter S_{11} specifies the reflection coefficient when both ports are terminated by impedance Z_o. Whereas for an amplifier, the termination at port 2 may not have much effect on the reflection at port 1, for devices such as transmission lines the reflection at port 1 is very much dependent on the termination impedance at port 2. Methods for calculating reflections and for converting two-port S-parameters into two-port impedance matrices can be found in most books on microwave engineering. Pozar (1998) is a great book covering a broad range of microwave techniques, including S-parameters and distributed circuited design. Artech House publishes an extensive series of microwave design texts. Good simulation software is a must for microwave engineering. Refer to the end of this chapter for a list of software companies and their web sites.

THE VECTOR NETWORK ANALYZER

The vector network analyzer (VNA) is the most commonly used instrument at high frequencies. The term "network analyzer" can be

Figure 10.8 The telephone line directional coupler (a.k.a., 2-wire/4-wire hybrid).

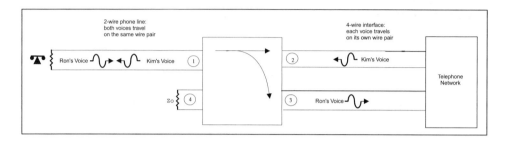

very misleading. In this context, network refers to an interconnection of electronic elements. In modern terminology, the term *network* usually refers to a computer or telephone network. However, when microwave engineering was developed (mostly in the 1940s), the term network was mostly used to describe "networks" of electronic components. It is important to keep in mind that a network analyzer has nothing to do with computer networks.

Now that I have explained what a VNA is not, I will explain what a VNA is. The VNA is an instrument that measures the S-parameters of 1-port and 2-port devices. Typically the S parameters can be displayed on a log-frequency scale or in a Smith chart format. Some VNAs can even convert reflection coefficients into their corresponding impedances. By the way, the term *vector* refers to the fact that a VNA measures both the amplitude and phase of the S-parameters.

The network analyzer works by sending a voltage wave into the device under test, and then measuring the reflected voltage wave to compute $\Gamma = S_1$. For a 2-port measurement, the VNA first applies a signal to port 1, then measures the reflected voltage wave at port 1 to compute S_{11} and measures the transmitted voltage wave at port 2 to compute S_{21}. The process is then repeated at port 2 to compute S_{22} and S_{12}.

Crucial to the operation of the network analyzer is a device called a *directional coupler*. This device acts like a multiplexer, in that it splits the single bidirectional port of the device into a separate input and output. Figure 10.6 should clarify this concept.

The directional coupler allows the VNA to apply a voltage wave and measure the returning reflection without the two waves being superimposed. At microwave frequencies, directional couplers are usually created from transmission lines or waveguides. Examples are shown in Figure 10.7.

Figure 10.9 An implementation of the telephone line directional coupler (a.k.a., 2-wire/4-wire hybrid).

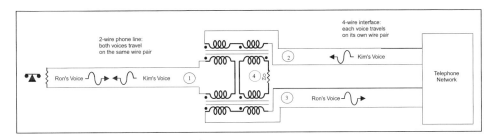

At lower frequencies, directional couplers can be created using transformers, LC circuits, or active amplifier circuits. The telephone companies actually use a directional coupler for voice signals from your telephone. These directional couplers are usually referred to as 2-wire/4-wire hybrids. A standard, plain old telephone service (POTS) telephone line is just a twisted pair of wires. The outgoing signal (your voice) and the incoming signal (the voice of the other person) are carried on the same pair of wires. The wave for each signal travels in a different direction. At the telephone company's side of the line, your voice signal is converted into a digital signal for transmission through the telephone network. In a reciprocal manner, the digitized signal of the other person is converted into an analog signal and sent down the phone line to your house. It is the hybrid that converts the 2-wire telephone line into a pair of 2-wire circuits (4-wire circuit)—one for the input signal going to the digital-to-analog-converter and one for the output signal coming from the analog-to-digital converter. Figure 10.8 shows the block diagram of the telephone line hybrid. Figure 10.9 shows a circuit implementation of the telephone hybrid.

BIBLIOGRAPHY

Surface Waves

Barlow, H. E. M., "Surface Waves: A Proposed Definition," *Proceedings of the IEE*, Vol. 107B, p. 240, 1960.

Barlow, H. M., and J. Brown, *Radio Surface Waves*, Oxford: Clarendon Press, 1962.

Barlow, H. M., and A. L. Cullen, "Surface Waves," *Proceedings of the IEE, Part III*, Vol. 100, pp. 329–347, 1953.

Boithias, L., *Radio Wave Propagation*, Boston: McGraw-Hill, 1987.

Brown, M., "The Types of Wave Which May Exist Near a Guiding Surface," *Proceedings of the IEE*, Part III, Vol. 100, p. 363, 1953.

Cullen, A. L., "The Excitation of Plane Surface Waves," *Proceedings of the IEE*, Part IV, Vol. 101, pp. 225–234, 1954.

Goubau, G., "Surface Waves and Their Application to Transmission Lines," *Journal of Applied Physics*, Vol. 21, November 1950.

Hill, D. A., and J. R. Wait, "Excitation of the Zenneck Surface Wave by a Vertical Aperture," *Radio Science*, Vol. 13, p. 969–977, 1978.

King, R. W. P., "The Wave Antenna for Transmission and Reception," *IEEE Transactions on Antennas and Propagation*, Vol. AP. 31, No. 6, November 1983.

King, R. W. P., M. Owens, T. T. Wu, *Lateral Electromagnetic Waves*, New York: Springer-Verlag, 1992.

Norton, K. A., "The Propagation of Radio Waves Over the Surface of the Earth and in the Upper Atmosphere—Part I," *Proceedings of the Institute of Radio Engineers*, Vol. 24, No. 10, October 1936.

Norton, K. A., "The Propagation of Radio Waves Over the Surface of the Earth and in the Upper Atmosphere—Part II," *Proceedings of the Institute of Radio Engineers*, Vol. 25, No. 9, September 1937.

Norton, K. A., "The Physical Reality of Space and Surface Waves in the Radiation Field of Radio Antennas," *Proceedings of the Institute of Radio Engineers*, Vol. 25, No. 9, September 1937.

Tamir, T., "Guided Complex Waves," *Proceedings of the IEE*, Vol. 110, No. 2, pp. 310–334. February 1963.

Wait, J. R., and D. A. Hill, "Excitation of the HF surface wave by vertical and horizontal antennas," *Radio Science*, vol 14, p 767–780, 1979.

Vector Network Analyzers and VNA Calibration

Agilent Technologies, *Specifying Calibration Standards for the Agilent 8510 Network Analyzer*, Agilent Technologies, PN 8510-5A.

Agilent Technologies, *Understanding the Fundamental Principles of Vector Network Analysis*, Agilent Technologies, AN 1287-1.

Agilent Technologies, *Exploring the Architectures of Network Analyzers*, Agilent Technologies, AN 1287-2.

Agilent Technologies, *Applying Error Correction to Network Analyzer Measurements*, Agilent Technologies, AN 1287-3.

Agilent Technologies, *Accessories Selection Guide for Impedance Measurements*, Agilent Technologies, April 2000.

Bauer, R. F., and P. Penfield, Jr., "De-Embedding and Unterminating," *IEEE Transactions on Microwave Theory and Techniques*, Vol. MTT-22, No. 3, March 1974.

Betts, L., "Calibrating Standards for In-Fixture Device Characterization," *Applied Microwave & Wireless*, November 2000.

Cascade Microtech, *On-Wafer Vector Network Analyzer Calibration and Measurements*, Oregon: Cascade Microtech Inc., 1997.

Engen, G. F., and C. A. Hoer, "'Thru-Reflect-Line': An Improved Technique for Calibrating the Dual Six-Port Automatic Network Analyzer," *IEEE Transactions Microwave Theory and Techniques*, Vol., MTT-27, No. 12, December 1979.

Gelnovatch, V. G., " A Computer Program for the Direct Calibration of Two-Port Reflectometers for Automated Microwave Measurements," *IEEE Transactions on Microwave Theory and Techniques*, January 1976.

Hewlett-Packard, *Effective Impedance Measurement Using OPEN/SHORT/LOAD Correction*, Hewlett-Packard, AN-346-3.

Hewlett-Packard, *Highly Accurate Evaluation of Chip Capacitors Using the HP 4291B*, Hewlett- Packard, AN-1300-1.

Hewlett-Packard, *Network Analysis: Applying the HP 8510 TRL Calibration for Noncoaxial Measurements*, Product Note: 8510-8A, Hewlett-Packard.

Soares, R. A., P. Gouzien, P. Legaud, G. Follot, "A Unified Mathematical Approach to Two-Port Calibration Techniques and Some Applications," *IEEE Transactions on Microwave Theory and Techniques*, Vol. 37, No. 11, November 1989.

Microwave Engineering and Distributed Circuits

Bigelow, S., *Telephone Repair Illustrated*, Blue Ridge Summit, PA: TAB Books, 1993.

Bigelow, S., *Understanding Telephone Electronics*, 3rd Edition, Boston: Newnes, 1997.

Bingman, G., "Transmission Lines of Antennas," *RF Design*, January 2000.

Caswell, W. E., "The Directional Coupler—1966," *IEEE Transactions on Microwave Theory and Techniques*, February 1967.

Cohn, S. B., and R. Levy, "History of Microwave Passive Components with Particular Attention to Directional Couplers," *IEEE Transactions on Microwave Theory and Techniques*, Vol. MTT-32, No. 9, September 1984.

Gupta, K. C., R. Garg, I. Bahl, P. Bhartia, *Microstrip Lines and Slotlines*, 2nd Edition, Boston: Artech House, 1996.

Johnson, H., "Strange Microstrip Modes," *EDN*, April 26, 2001.

Matthaei, G., E. M. T. Jones, L. Young, *Microwave Filters, Impedance-Matching Networks, and Coupling Structures*, Boston: Artech House, 1980.

Mongia, R., I. J. Bahl, P. Bhartia, *RF and Microwave Coupled Line Circuits*, Boston: Artech House, 1999.

Monteath, G. D., "Coupled Transmission Lines as Symmetrical Directional Couplers," *Proceedings of the IEEE*, No. 1833 R, May 1955.

Oliner, A. A., "Historical Perspectives on Microwave Field Theory," *IEEE Transactions on Microwave Theory and Techniques*, Vol. MTT-32, No. 9, September 1984.

Oliver, B. M., "Directional Electromagnetic Couplers," *Proceedings of the IRE*, November 1954.

O'Rourke, R., "Designing Microwave Circuits with Electromagnetic Simulation," *RF Design*, May 1998.

Parisi, S. J., "A Lumped Element Rat Race Coupler," *Applied Microwave*, 1989.

Philips Semiconductors, *Circulators and Isolators, Unique Passive Devices*, Philips Semiconductor, AN-98035.

Pozar, D. M., *Microwave Engineering*, 2nd Edition, New York: John Wiley, 1998.

Ramo, S., J. R. Whinnery, T. Van Duzer, *Fields and Waves in Communication Electronics*, 2nd Edition, New York: John Wiley, 1989.

Sevick, J., "Notes on Power Combiner and Splitter Circuits," *Applied Microwave & Wireless*, February 1999.

Shifrin, M., C. Lyons, W. Grammer, P. Katzin, "An Electronic Directional Coupler," *Applied Microwave & Wireless*, November 1996.

Wadell, B. C., *Transmission Line Design Handbook*, Boston: Artech House, 1991.

RF, Microwave Circuit, and Electromagnetic Simulation Software

Agilent EESOF, http://contact.tm.agilent.com/tmo/hpeesof/
Applied Wave Research, http://www.appwave.com/products/index.html
Ansoft, http://www.ansoft.com/
Cosmos, http://www.cosmosm.com/
Eagleware GENESYS, http://www.eagleware.com/home.html
Remcom, http://www.remcom.com/html/index.html
Vector Fields, http://www.vectorfields.com/
Zeland, http://www.zeland.com/

Web Resource

http://sss-mag.com/spara.html#tut

11 ANTENNAS: HOW TO MAKE CIRCUITS THAT RADIATE

In Chapter 5, I discussed how and why circuits radiate, without any regard to practical implementation of antennas. This chapter covers how to make circuits that radiate efficiently. In Chapter 5, I based the results on an ideal line of current. In practice, circuits usually have a return current that flows in the opposite direction of the signal current. For example, in a two-wire transmission line, the two wires are parallel with signal current flowing in one direction and return current flowing in the opposite direction. There is a field between the wires and in the immediate vicinity of the wires, but at far distances, the fields from the two wires tend to cancel one another. The result is that very little radiation occurs. There are two basic ways to create an antenna from a transmission line. One method results in an electric dipole and the other method results in a magnetic dipole.

THE ELECTRIC DIPOLE

To create a half-wavelength dipole, start with an open-circuit transmission line of length equal to $\lambda/4$, as shown in Figure 11.1. Applying a voltage to the line creates a standing wave, and the input impedance appears (ideally) as a short circuit. With this geometry, very little radiation occurs. Now imagine that you take the ends of the transmission line and expand the wires outward 90 degrees. A standing wave still occurs on the line, with current going to zero at the ends. However, now the current in the wires travels in the same direction, producing a very efficient radiator, a $\lambda/2$ dipole. Since power is continually propagated from the antenna, the input resistance is no longer zero. There is now a resistance of (ideally) $R_{input} = 73$ ohms, and the power transmitted is

$$P = V \times I = I^2 \times R_{input}$$

Since the antenna has an open-circuit form (high-impedance), the near field is dominated by the electric field.

Figure 11.1 A dipole antenna can be created by pulling apart an open-circuit transmission line. The arrows denote the direction of current, and the dotted lines indicate current magnitude.

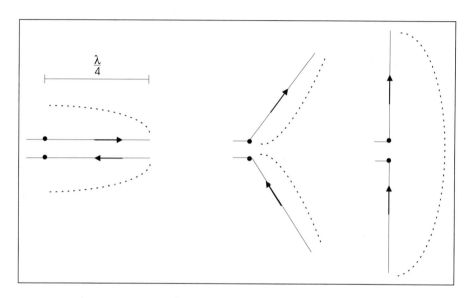

THE ELECTRIC MONOPOLE

A monopole antenna is half of a dipole antenna. If a monopole is placed vertically above a conductor, such as the Earth, the conductor acts as a reflector and creates an image of the other half of the dipole. For a monopole to work most efficiently, the circuit ground should be connected to the conductor used for reflecting. The radiation pattern of the monopole above the ground has the same shape as the upper half of the pattern of a vertical dipole. Furthermore, the radiation resistance is exactly one half the value of a dipole of twice the length of the monopole. For example, a $\lambda/4$ monopole has half the radiation resistance of a $\lambda/2$ dipole:

$$R_{\lambda/4\,monopole} = 0.5 \times R_{\lambda/2\,dipole} = 73/2 = 36.5 \,\text{ohms}$$

THE MAGNETIC DIPOLE

The twin to the electric dipole is the magnetic dipole. To create a full-wavelength magnetic dipole, start with a short-circuited transmission

Figure 11.2 A loop antenna can be created by pulling apart a short-circuit transmission line. The arrows denote the direction of current, and the dotted lines indicate current magnitude.

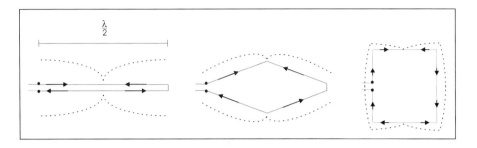

line of length equal to $\lambda/2$. Applying a voltage to the line creates a standing wave, and the input impedance appears (ideally) as a short circuit. The next step is to pull the wires apart to create the maximum area between them, as shown in Figure 11.2. The maximum area is produced when the wires are pulled apart to form a circle, although square loops also work well. The currents in the wire no longer cancel each other. Instead, the current flows in a circular motion. Since the antenna forms a short circuit, in the near field the field is dominated by the magnetic field. The magnetic dipole is commonly called a loop antenna. The radiation resistance of the full-wavelength is (ideally) about 100 ohms.

RECEIVING ANTENNAS AND RECIPROCITY

An important property of antennas is the law of reciprocity. The law of reciprocity states that the transmission and reception properties of an antenna are equal. An antenna that transmits efficiently will also receive efficiently. Characteristics such as antenna pattern and feeding impedance are the same regardless of whether the antenna is in a transmitting or receiving circuit. Because of this law, two-way devices such as cell phones can use the same antenna for transmitting as well as for receiving.

RADIATION RESISTANCE OF DIPOLE ANTENNAS

As I discussed in Chapter 5, the amount of power that an antenna radiates is a function of the electrical length of the antenna. A 1 MHz

(AM radio) electric dipole antenna that is 150 m long radiates the same amount of power as a 100 MHz (FM radio) electric dipole that is 1.5 m long. The equivalence stems from the fact that both antennas have the same electrical length, $\lambda/2$. The radiating efficiency is usually expressed in terms of a quantity called *radiation resistance*. The power emitted by an antenna can be calculated by taking the product of the radiation resistance and the average current on the antenna,

$$P = I^2_{antenna} \times R_{radiation}$$

It is important to keep in mind that the current on antennas is not necessarily the same at different points along the antenna. Just as with long, unmatched transmission lines, the current on antennas can form standing waves. For example, the current on a dipole must go to zero at the ends because the ends are unconnected.

For electrically small antennas, the radiation resistance of an electric dipole is proportional to the second power of the electrical length of the antenna.

The radiating efficiency from magnetic dipole antennas is also a function of electrical size of the loop. For a loop antenna, the radiation resistance depends on the area of the loop, and is therefore proportional to the fourth power of the electrical length of the loop circumference. Figures 11.3 and 11.4 compare the radiation resistance of electric and magnetic dipoles as a function of electrical size.

FEEDING IMPEDANCE AND ANTENNA MATCHING

Just as the input impedance of a transmission line depends on length, so does the input impedance of an antenna. The input impedance that an antenna presents to the amplifier or transmission line feed is not necessarily equal to the radiation resistance.

An electric dipole behaves like an open-circuited transmission line, in that it has a capacitive input impedance when its length is electrically short ($\ell < \lambda/2$). At resonance ($\ell \sim \lambda/2$), its impedance is purely real, with the resistance relating to the radiated power. The resonant point occurs at a length slightly less than $\lambda/2$ because of end effects—the ends of the antenna have parasitic effects that cause the antenna to act slightly larger than it is. A decent rule of thumb is to design an antenna about 2% to 5% smaller in size for the electrical length desired. Above resonance, the electric dipole has an inductive input impedance. The impedance is purely real at every $\lambda/2$, and switches between capacitive

Figure 11.3 Radiation resistance versus electric dipole length. Below a half wavelength, the radiation resistance is proportional to L_λ^2 (the square of the electrical length). Power radiated can be determined using $P = I^2 \times R_{rad}$.

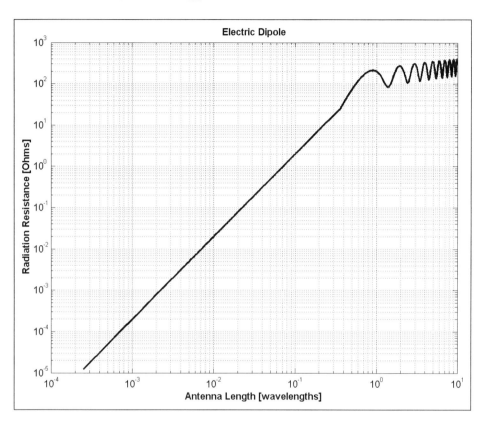

and inductive impedance on either side of these nodes. In addition, the real portion of the feeding impedance does not always equal the radiation resistance. The radiation resistance is transformed similarly to the way a transmission line transforms impedance.

The law of maximum power transfer (Chapter 7) requires that the antenna must be conjugate matched to the feeding transmission line (or amplifier if the line is electrically short) for maximum power to be exchanged between the amplifier and the antenna. In other words, the reactive portion of the antenna impedance must be cancelled with equal and opposite reactance and the real portion should be converted to the transmission-line/amplifier impedance using an impedance

Figure 11.4 Radiation resistance versus magnetic dipole circumference. Below a wavelength, the radiation resistance is proportional to C_λ^4. Power radiated can be determined using $P = I^2 \times R_{rad}$. This plot assumes a constant current around the loop, and loses accuracy above a half wavelength. However, the general behavior of large loop antennas is similar.

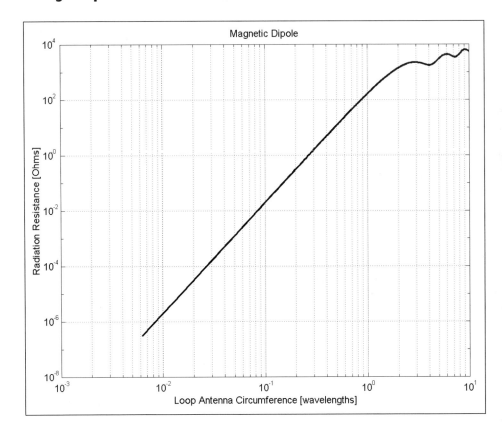

matching network. The large input impedance at multiples of λ makes these lengths hard to match.

The magnetic dipole behaves like a short-circuited transmission line. When its length is electrically short, it exhibits an inductive input impedance. At resonant length (circumference = $\lambda/2$), the antenna exhibits a purely real input impedance. Similar to the shorted $\lambda/4$ transmission line, this impedance is large, typically 2 kohm to 5 kohm. Above resonance, the input impedance is capacitive. At the first harmonic resonance (circumference = λ), the input impedance is real again, and then changes back to inductive impedance when the length is slightly larger

Figure 11.5 Feed (input terminal) impedance of an electric dipole as a function of antenna length. For reference, the radiation resistance is shown as a solid line. At a half wavelength, the feed impedance is equal to the radiation resistance. At a wavelength, the feed impedance has a large inductive reactance and the feed resistance is substantially larger that the radiation resistance. Feed impedance is also highly dependent on the wire radius of the antenna. The plots show feed resistance for two different ratios of length to radius.

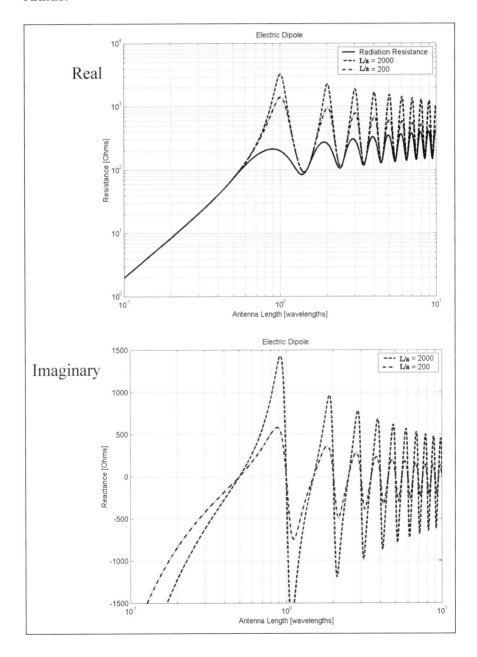

Table 11.1 Calculated Characteristics of Loop Antennas of Various Sizes

Circumference	Feed Resistance	Feed Reactance	Radiation Resistance
0.1λ	$0.80\,\Omega$	$+i\,110\,\Omega$ (inductive)	$0.020\,\Omega$
0.5λ	$5000\,\Omega$	$-i\,2500\,\Omega$ (capacitive)	$12\,\Omega$
1.0λ	$106\,\Omega$	$-i\,90\,\Omega$ (capacitive)	$117\,\Omega$
1.14λ	$150\,\Omega$	$i\,0\,\Omega$ (resonant)	$125\,\Omega$

Ratio of loop circumference to wire radius was 950:1. Wire was assumed to be perfectly conducting.

than λ. The full-wavelength magnetic dipole has a low feed impedance of about 100 ohms, about equal to the radiation resistance. For this reason, large magnetic loops are usually designed with the circumference equal to λ instead of $\lambda/2$. The loop antenna is not exactly resonant at $C = \lambda$, but at a slightly longer length, typically 5% to 10% longer than a wavelength. The feed impedance of electrically large magnetic loops is extremely difficult to calculate because of a complex standing wave that exists on the loop. Plots are given in Johnson (1993). The mathematical theory can be found in King and Harrison (1969), Collin and Zucker (1969), and Lo and Lee (1988). The feed impedance is also very dependent on the ratio of loop circumference to the wire radius of the wire gauge used in the loop. Example feed impedances are given in Table 11.1.

ANTENNA PATTERN VERSUS ELECTRICAL LENGTH

Antennas do not radiate equally to all directions. The directional dependence of antenna radiation is aptly called the *antenna pattern*. An electric dipole that is $1/2\,\lambda$ in length or smaller exhibits a fairly simple antenna pattern. The most power is radiated in the direction broadside to the antenna; that is, perpendicular to the antenna's length. The power decreases with decreasing angle such that at zero degrees, no power is radiated. In other words, no power is radiated in the direction of the end points. This behavior is fundamental to charges moving along a line. Referring to Figures 5.5D and 5.6B, you can see that no power is radiated in the direction that the charge moves. Figure 11.6 shows the three-dimensional plot of the radiated field power of a half-wavelength dipole.

Figure 11.6 The 3D radiation pattern of a half-wavelength dipole antenna, shown in both horizontal and vertical orientations.

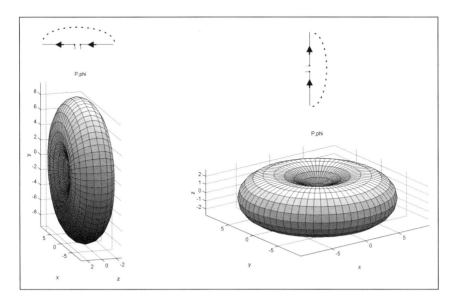

In small antennas, the current goes to zero only at the end points. In large antennas with electrical length greater than λ, other points of zero current appear. These locations of zero current that occur at intermediate distances along the antenna are standing wave nodes like those found on transmission lines. These nodes of zero current correspond to directions in which no radiation is received. The resulting antenna has regions, called *lobes*, of reception, which are separated by the nodes. Figure 11.7 shows the antenna pattern for electric dipoles of various lengths.

The magnetic dipole also exhibits an antenna pattern that is dependent on its electrical length. At small electrical lengths the majority of the power is radiated around the axis of symmetry. To understand what the axis of symmetry is, imagine a bicycle wheel. The axis of symmetry is the axle of the wheel. The radiation from the antenna is most prominent in the direction of the spokes that emanate from the axle. For electrically small loops, the radiation is zero in the direction of the axis of symmetry (in the direction of the axle).

As the loop is increased, a standing wave of current is created around the loop. The standing wave is quite complicated in form and causes the radiation pattern to lose its symmetry. The other interesting effect

Figure 11.7 Radiation patterns for electric dipoles of various electrical lengths. In each case, the antenna is driven with 1 A. The current on the antenna is shown for lengths of 1/2 and 3/2. The center plot shows the end-view pattern for all lengths.

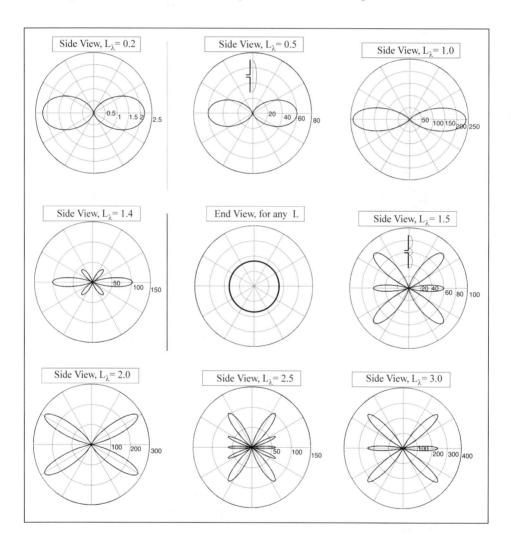

of the standing wave is that it causes radiation to be emitted broadside to the loop (in the direction of the loop's axis of symmetry). Figure 11.8 shows the three-dimensional radiation patterns of loops of three different electrical sizes. Finally, the radiation from loop antennas contains waves of different polarizations.

POLARIZATION

Figure 11.8 Radiation patterns for magnetic loop antennas of three electrical sizes (given in terms of circumference). The top row shows the antenna with feed from each view.

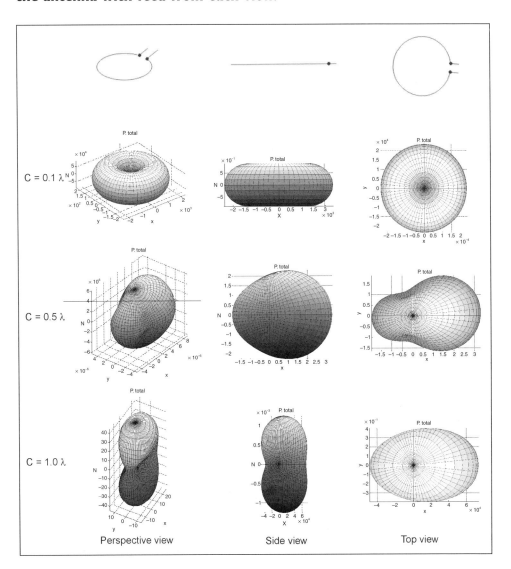

POLARIZATION

When an electromagnetic field travels in space, its electric field and magnetic field components are transverse to the direction of propagation and are at 90 degree angles to one another. Even with these limitations, the field components have 360 degrees of freedom in terms of which direction the fields point. The antenna geometry determines the direction of the fields. For an electric dipole, the electric field will be polarized in the direction of the dipole wire. The magnetic field is polarized at a right angle to the electric field. Magnetic dipole antennas polarize the magnetic field in the direction created by the axis of symmetry.

For antennas on Earth, antenna polarization is usually described as either horizontal or vertical in reference to the surface of the Earth. For instance, the monopole radio antenna of a car is vertically polarized, but dipole antennas like those typically used on indoor FM radio receivers are typically mounted horizontal to the ground, implying horizontal polarization. The broadcast antenna may or may not be of the same polarization as the receiving antenna. If the polarization of the transmitted wave and the receiving antenna is opposite, 3 dB of signal power is lost. Thus, for maximum reception, you should orient your receiving antenna to match the polarization of the transmitted wave. AM radio is always vertically polarized because large monopole towers are used. In the case of AM antennas, the wavelengths are so long that the tower is actually the antenna. In other words, the entire tower is electrified and acts as a giant monopole antenna. FM and TV broadcasts can be vertically, horizontally, or circularly polarized. A circularly polarized wave contains both vertical and horizontal polarized waves in equal proportions. The wave produced is called circularly polarized because the polarization rotates as the wave propagates. Circularly polarized waves can be received equally well by both vertical and horizontal dipole antennas.

The Indoor FM Radio Antenna

Horizontal mounting of FM-band indoor antennas is most often suggested because the reflectivity of the ground is typically better for horizontally polarized waves than it is for vertically polarized waves (see Chapter 15). Thus, at far distances from the broadcast antenna, the horizontally polarized wave has greater power than the vertically polarized wave. However, there is one major problem with horizontally mounting an FM dipole antenna, and that is the antenna's pattern (refer back to Figure 11.6). When you horizontally mount an FM antenna on the wall inside your home, the antenna will properly receive signals from directions perpendicular to the wall, but will receive no signals from directions parallel to the wall! The result is that you get great reception for signals from certain directions and terrible reception for signals from other directions. Since you cannot rotate the direction of your living room wall, you have to be very lucky to have all your stations in directions perpendicular to the wall mounting. Another option, which I have found to work quite well, is to mount the FM dipole vertically on your wall. Be sure to place the vertical mounting $\lambda/3$ (~3 feet) or higher above the ground to limit ground effects (described in the following section). Of course, a better solution would be a rotating horizontal dipole antenna, which could be directed toward the broadcast station. A Yagi-Uda antenna, described later, is even better.

EFFECTS OF GROUND ON DIPOLES

For most antenna applications, the antenna is placed within the vicinity of the Earth's surface. Because the Earth is a conductor, the antenna pattern is affected by the presence of the Earth. A monopole antenna is designed to be placed over a ground, but dipole antennas are designed to operate in free space. The Earth affects the feeding impedance, radiation resistance, and the antenna pattern of dipole antennas. The effects are greatest when the antenna is close to the ground, and gradually diminish as the antenna height above the ground is increased. Figures 11.9 and 11.10 illustrate some of the ground effects on dipole antennas. The figures assume a perfectly reflecting ground. Any real ground is, of course, far from being a perfect conductor. Furthermore, the ground resistance varies from place to place and also varies with frequency. Hutchinson, Kleinman, and Straw (2001) contains a map of the typical ground conductivity as it varies across the United States. Refer to Chapter 15 for the frequency variation of a typical ground.

The $\lambda/4$ Monopole Antenna

The vertical monopole antenna or "whip" antenna is a commonly used alternative to the dipole antenna. It consists of a single wire that extends from a coaxial transmission line. The signal wire is connected to the monopole wire, and the ground shield is left unconnected or is connected to ground. When the antenna is mounted within close electrical distance to the ground, the $\lambda/4$ monopole acts very like the $\lambda/2$ dipole. This antenna length is resonant, although it is slightly less efficient than the $\lambda/2$ dipole. The ideal vertical monopole has a radiation resistance of 36.5 ohms, as compared to 73 ohms of the dipole antenna. To ensure good performance of the monopole antenna, a metal plate or metal mesh can be placed below the monopole, creating a good conducting reference. The ground wire of the driving circuit is then

Figure 11.9 Feed/radiation resistance and feed capacitance for λ/2 electric dipoles versus height above ground.

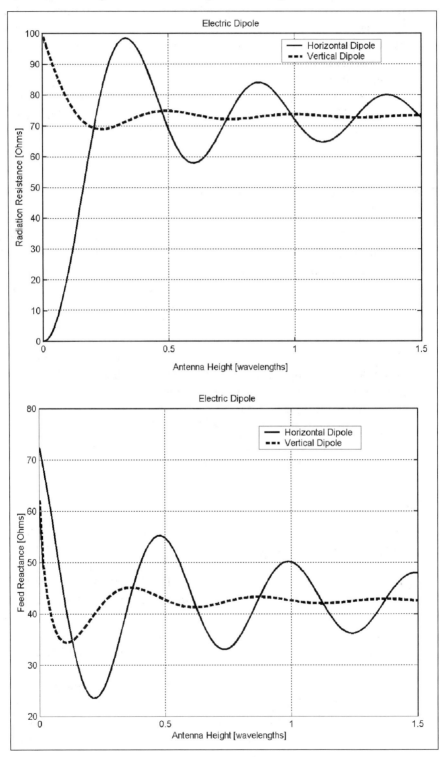

Figure 11.10 Antenna patterns for a λ/2 horizontal electric dipole at various heights above a perfect ground. Heights are given in terms of wavelength. The antenna was assumed to have Ohmic resistance of 5 ohms, corresponding to heat loss in the antenna.

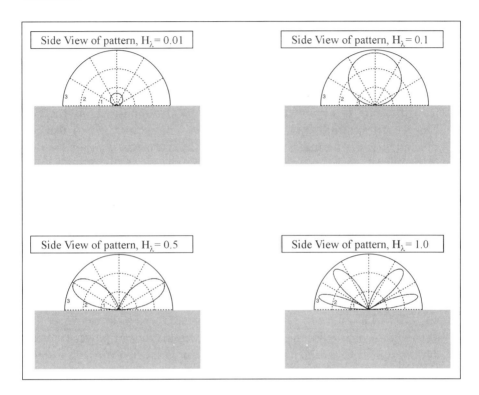

connected to the plate. Another method for virtual ground referencing is the use of *ground radials*, also known as *ground counterpoise*, which are electrically long wires that extend radially from the base of the antenna. As an alternative to using long radials, several λ/4 radials can be used and pointed downward at an angle of 45 degrees. By angling the radial downward the radiation resistance is increased to about 50 ohms, making the antenna more efficient and easy to match to standard 50 ohm coaxial cable.

The 5/8 λ Monopole Antenna

The 5/8 λ monopole antenna is used for antenna applications where the goal is transmission/reception of maximum power in the direction broadside (perpendicular) to the antenna. The 5/8 λ monopole antenna has an antenna pattern with a single lobe, which is fairly narrow and concentrates the antenna signal. So, although it doesn't produce largest total power, it is an efficient radiator that delivers the highest percentage of power in the broadside direction. The exact theoretical value for this condition is $2/\pi \times \lambda = 0.64\ \lambda$, but the approximation of $5/8 = 0.625\ \lambda$ is most often used.

The Car Radio Antenna

The antenna on most cars is used for both AM and FM band reception. This antenna is usually a monopole vertically mounted on the car chassis. The ground lead of the car antenna is connected to the car chassis at the antenna. The chassis acts like a virtual ground for the antenna, allowing for good reception. The length of the monopole is typically about 75 cm, which is a quarter wavelength at the center of the FM band (87.9 MHz to 108.1 MHz). The same antenna is also used for AM radio reception. The AM band spans the frequency range of 0.540 MHz to 1.610 MHz. At the center of the AM band, the 75 cm antenna has an electrical length of about 0.004 wavelengths. The relative power that it receives is about 35 dB to 40 dB less than that of the antenna in the FM band. In addition, the antenna has a capacitive feed impedance at the AM frequency range. Therefore, the AM receiver includes an inductor in series with the antenna feed so as to cancel the capacitance of the antenna and provide the maximum power transfer from antenna to receiver. The signal is then connected through a step-down transformer, which reduces the effective input impedance of the electrically short antenna, also allowing for greater power transfer.

WIRE LOSSES

When current is carried by a transmission line, some power is lost in the wires of the transmission line in addition to that delivered to the load. With antennas, the current that creates the radiation also causes resistive losses in the antenna itself. The resistive losses limit the efficiency of the antenna, causing some of the input power to be lost as heat. The way to reduce the resistive losses is to use thicker wire for the antenna. The thicker wire has less resistance, leading to smaller Ohmic losses. From Figure 11.5, you can see that the wire radius transforms the feed impedance of the antenna. The thickness of the antenna wire also has an effect on the radiating behavior of the antenna. For a

ANTENNAS, ANTENNA APERTURE, AND RADAR CROSS-SECTION   245

dipole, increasing the wire thickness also reduces the radiation resistance very slightly. In summary, the wire radius contributes three effects: it adds heat losses, changes the feed impedance, and changes the radiation resistance.

SCATTERING BY ANTENNAS, ANTENNA APERTURE, AND RADAR CROSS-SECTION

A receiving antenna captures a certain amount of the radiation incident in its vicinity, but how much? The most common method for defining the amount of radiation an antenna intercepts is the use of effective cross-section. In this method, you calculate the effective area of the antenna as if the incident radiation consists of rays, like the rays of geometric optics. For rays of light, the cross-sectional area is simply equal to its area. The concept is familiar to us. If a distant light is shown on an opaque object, the region behind the object will be in shadow. The cross-sectional area of the shadow is the same as the cross-sectional area of the object. Since the width of the dipole antenna is always much smaller than a wavelength, the methods of geometrical optics cannot be used. However, the amount of radiation captured by an antenna can be calculated or measured. From the calculated/measured value, an equivalent cross-section can be determined. For example, a simple $\lambda/2$ electric dipole, terminated into a matching impedance, has an effective cross-sectional area of $0.13\ \lambda^2$. This area can be approximated by a $\lambda/2$ by $\lambda/4$ rectangle or by an ellipse with major axis diameter of $\lambda/2$ and minor axis diameter of $0.34\ \lambda$. This area relates to the amount of power absorbed in the load, presumably the receiving amplifier. The $\lambda/2$ electric dipole also scatters or reradiates an equal amount of radiation. Now consider an electric dipole of arbitrary length (ℓ), terminated in a matched load with $\ell \ll \lambda/2$. Its effective cross-section is approximately $0.119\ \lambda^2$, and it also reradiates an amount equal to what it absorbs. What about dipoles that are not connected to a matched load? If a short is placed across the leads of a $\lambda/2$ dipole, then no power is delivered to the load, but the antenna collects radiation with a (scattering) cross-section of $0.52\ \lambda^2$ and reradiates all of this power. The shorted dipole is equivalent to an unconnected metal rod of length $\lambda/2$. *This concept is important to understand; a conducting rod by itself in space, without any connecting circuitry, can collect and reradiate a large amount of power.* This energy is large when the rod is at a resonant length, and very small when at a nonresonant length. The scattering cross-section of objects is closely related to *radar cross-section*, which quantifies the amount of radiation reflected from a source "illuminated" by a radar pulse.

DIRECTED ANTENNAS AND THE YAGI-UDA ARRAY

Antenna elements without connecting wires are used in certain multi-element antennas, and are called passive elements. As you have just learned, such passive antennas can capture and re-emit radiation. When passive elements are placed within the near-field of an active antenna, the behavior of the active antenna can be enhanced.

Consider a $\lambda/2$ dipole with a passive rod placed to the side of it as shown in Figure 11.11. If the passive rod is placed in the near field, coupling between the active and passive elements will result. If the passive element is slightly larger than the active element, it will act like a reflector. If the passive element is slightly smaller than the active element, it will act as a director. In both cases, the pattern of the antenna is modified so that there is gain in a specific direction, as shown by the figure.

To produce even better results, you can construct an antenna consisting of a $\lambda/2$ dipole, with both a reflector and director. Such an antenna is called a Yagi-Uda antenna, named after the two Japanese researchers who invented it in the 1920s. Yagi-Uda antennas and their variants are used very often as receiving antennas, because you can point the antenna in the direction of the transmitter to receive several dB of gain in signal strength over a simple dipole. In addition to the gain achieved, unwanted signals are attenuated because the antenna pattern shrinks in directions other than the direction to which it is pointed. Yagi-Uda antennas are often used for both transmitting and receiving in fixed location radio systems, such as weather stations. Because the position of the transmitting antenna at the weather station, and the position of the receiving antenna where the data is analyzed, never change, a Yagi-Uda antenna provides an easy method for improving signal strength.

TRAVELING WAVE ANTENNAS

All of the electric antennas considered so far are of the standing wave variety. Each of these antennas is conceptually created by opening up an open-circuit antenna. The open-circuit characteristic causes the antenna to be unmatched, leading to the standing waves. If an antenna is matched at one end, a traveling wave antenna is created. The result

TRAVELING WAVE ANTENNAS 247

Figure 11.11 Parasitic antenna elements can be used to direct antenna radiation and increase gain.

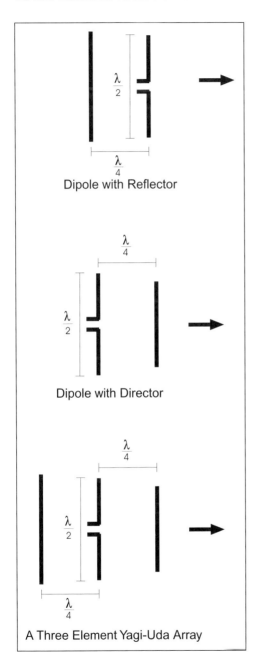

is that the entire antenna pattern is bent toward the terminated end of the antenna. A common use of this phenomenon is the use of an electrically long monopole antenna terminated via a resistor connected to the Earth. Traveling wave antennas are sometimes called nonresonant or terminated antennas.

ANTENNAS IN PARALLEL AND THE FOLDED DIPOLE

Electric dipole antennas can be placed in parallel to produce more radiation. In fact, two electric dipoles placed in close proximity create a near field coupling that produces radiation power that is a factor N^2 greater than a single dipole, where N is the number of dipoles.

The folded dipole, shown is Figure 11.12, is another common and simple antenna. At first glance, the folded dipole looks like a transmission line, and therefore does not seems to be a good radiator. However, the key to the folded dipole is the feeding of the antenna. It is fed at a break in the one line at the center of the structure. Another way of looking at it is as a loop, with a circumference equal to lambda, that has been squeezed into a slender form. At low frequencies, where the folded dipole is electrically small, this antenna is a very poor radiator because the current in the parallel wires tends to cancel each other as in a transmission line.

Figure 11.12 A folded dipole antenna. The arrows denote the direction of current.

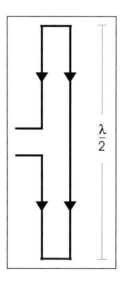

When the antenna is electrically large, the currents in the two parallel wires become in phase. At resonance, this antenna acts like two closely spaced dipole antennas, quadrupling the total radiation. The novel feed connection thus allows the transmission line to behave like two parallel dipoles in the neighborhood of resonance. It has an antenna pattern similar to a standard electric dipole, but radiates more efficiently with a radiation and feed resistance of $R_{radiation} = R_{dipole} \times N^2 = 73 \times 2^2 = 292$ ohms. The folded dipole is often used as an indoor VHF TV antenna.

MULTITURN LOOP ANTENNAS

I have focused on electric antennas, because these antennas are used most often in practice. One reason is that loops tend to take up more space than electric dipoles. However, the loop antenna can actually save space in low-frequency radio applications, where wavelength is tremendous. For instance, the wavelength at the center of the AM radio band is about 300 m. Creating an electrically long antenna at this wavelength is not feasible, except for the broadcast companies. An alternative to the use of an electrically short monopole, like that of the car radio, is the use of a multiturn loop antenna. Multiturn loop antennas have the useful property that radiation power increases with N^2, where N is the number of turns in the antenna. In other words, the radiation resistance of a multiturn loop antenna is $N^2 \times R_{loop}$, where R_{loop} is the radiation resistance of a single loop. The electrically short multiloop antenna is commonly used in portable radios for AM reception. Most portable AM/FM radios have two antennas, a telescoping monopole antenna for the FM band and an internal multiturn loop antenna for the AM band. Therefore, if you wish to improve the reception of an AM radio station, fiddling with the telescoping monopole will not do you any good. On the contrary, you should rotate the orientation of the whole radio when looking to improve AM reception! By rotating the whole radio, you are rotating the internal AM loop antenna. The AM antenna usually consists of a many turn ($N \sim 100$) loop of wire around a ferromagnetic or ferrite rod. The ferrite rod serves to concentrate the local magnetic field to enhance the reception by a factor μ_r, where μ_r is the relative permeability of the ferrite. Because the AM band frequencies are relatively low (~1 MHz), the ferrite is low-loss. Such antennas are commonly called loop stick antennas, because they consist of many loops of wire around a ferrite stick.

There are many other types of antennas, including broadband antennas such as those used for roof-mounted TV reception, but with an understanding of the basics of antenna operation, these other more sophisticated designs can be easily understood. Kraus (1988) is a great theoretical book on antennas, considered the bible of antenna design. King and Harrison (1969) covers advanced antenna theory in the most complete manner. Straw (2000) and Carr (2001) are useful books covering practical antenna design and antenna construction aimed at amateur radio applications.

BIBLIOGRAPHY AND SUGGESTIONS FOR FURTHER READING

Balanis, C., *Antenna Theory: Analysis and Design*, New York: Harper & Row, 1982.

Bingman, G., "Transmission Lines of Antennas," *RF Design*, January 2000.

Carr, J. J., *Joe Carr's Receiving Antenna Handbook*, Solanna Beach, Calif.: HighText Publications, 1993.

Carr, J. J., *Practical Antenna Handbook*, Fourth Edition, New York: McGraw-Hill, 2001.

Collin, R. E., and F. J. Zucker, *Antenna Theory Part 1 and Part 2*, New York: McGraw-Hill, 1969.

Elliot, R. S., *Antenna Theory and Design*, Englewood Cliffs, NJ: Prentice-Hall, 1981.

Hutchinson, C., J. Kleinman, and D. R. Straw, Editors, *The ARRL Handbook for Radio Amateurs*, 78th Edition, Newington, Conn.: American Radio Relay League, 2001.

Johnson, R. C., and H. Jasik, Editors, *Antenna Engineering Handbook*, New York: McGraw-Hill, 1993.

Jordan, E. C., and K. G. Balmain, *Electromagnetic Waves and Radiating Systems*, 2nd Edition, Englewood Cliffs, NJ: Prentice Hall, 1968.

King, R. W. P., and C. W. Harrison, *Antennas and Waves: A Modern Approach*, Boston: The M.I.T. Press, 1969.

Kraus, J. D., *Antennas*, 2nd Edition, Boston: McGraw-Hill, 1988.

Lo, Y. T., and S. W. Lee, editors, *Antenna Handbook—Theory, Applications, and Design*. Van Nostrand Reinhold Company, 1988.

Orr, W., *Radio Handbook*, 32nd Edition, Woburn, Mass.: Butterworth-Heinemann, 1997.

Paul, C., and S. Naser, *Introduction to Electromagnetic Fields*, 2nd Edition, Boston: McGraw-Hill, 1987.

Ramo, S., J. R. Whinnery, and T. Van Duzer, *Fields and Waves in Communication Electronics*, 2nd Edition, New York: John Wiley, 1989.

Straw, R. D., Editor, *The ARRL Antenna Book*, 19th Edition, Newington, Conn.: American Radio Relay League, 2000.

12 EMC

PART I: BASICS

Designing circuits that meet product requirements is the primary goal of electronics. Designing circuits that can function in the real-world environment of radio interference, static electricity, lightning, power-line (60 Hz) interference, and so on without failure or damage is a secondary but equally important goal. Not only do products have to withstand the interference of other circuitry, they are also limited by law in the amount of interference they are allowed create. The study and resolution of such problems falls under the discipline of electromagnetic compatibility (EMC). In the past, interference problems were often referred to as radio-frequency interference (RFI) or electromagnetic interference (EMI). EMC is the modern term used to describe the need to both withstand interference and limit the production of interference, which is emphasized by the word "compatibility." The discipline of EMC also includes analyzing electrostatic discharge (ESD) and lightning problems. Products need to be compatible with their environment. In the United States, the Federal Communications Commission (FCC) regulates EMC. In most of Europe, EMC requirements are now regulated by the European Union (EU). Other regulations include those of the International Committee on Radio Interference (CISPR Publication 22) and the U.S. military standards (MIL-STD-461).

SELF-COMPATIBILITY AND SIGNAL INTEGRITY

A related issue to EMC is that of *self-compatibility*, or making sure that different circuits within a product do not interfere with one another. There are no laws governing self-compatibility, but without it your product may not function as intended. In digital systems, the term *signal integrity* is used to describe the analog aspects of digital signals.

Digital signals are ideally square waves, but in high-speed systems, they can be quite different from the ideal.

Real digital signals are limited by finite rise times. Oscillations, often called ringing, can occur due to transmission line reflections and parasitic reactance in components, including the transistors themselves. Other common effects are overshoot/undershoot and shelves (nonmonotonic behavior). Signal integrity of digital signals can be adversely effected by poor layout of the printed circuit board, by lack of proper handling of transmission line effects, by cross-coupling from other signals, or by overloading a driving circuit. In the high-speed digital systems of today's world, digital designers must have some understanding of radio-frequency effects and of how layout of the circuit can affect the digital signals. When the signal integrity of digital signals becomes poor, the signal can be read improperly at the input, producing signal glitches. Digital logic is based on thresholds that determine which state the signal is in (high or low). Effects such as ringing can cause a signal to be misinterpreted by the input logic. At this point the analog problem has become a digital problem. Figure 12.1 shows various aspects of high-speed digital signals and how errors can arise.

FREQUENCY SPECTRUM OF DIGITAL SIGNALS

The frequency spectrum of digital signals is probably the most important topic in understanding signal integrity for digital systems. As I have mentioned throughout this book, higher-frequency signals produce higher radiation. *The largest contributing factor to the high-frequency content of digital signals is the rise time (and fall time)*. Rise time has a much greater effect than does how often the signal changes (e.g., clock frequency). Rise times are the real culprit of most digital problems, not the clock frequency. The frequency spectrum of digital signals is broad in range. For a clock (a repetitive digital signal of exact period/frequency), the spectrum (see Figure 12.2) consists of a spike at the fundamental frequency of the clock and at all the odd harmonics of this frequency. This spectrum contains high-powered harmonics up to the equivalent frequency of the rise time,

$$f_{rise\,time} = 0.34/t_{rise}$$

Cross-coupling between two parallel signal traces is a common problem in digital systems. While this is a near-field effect (as opposed to a radiation effect), it also increases with frequency, as I will describe shortly. When the layout is poorly designed, the near field of signals can

Figure 12.1 Examples of digital signaling anomalies. Digital devices have three distinct input voltage regions: low, high, and transition. The output is undefined while the input is in the transition region. Any nonmonotonic behavior in the transition region can cause output glitches. Nonmonotonic behavior in the transition region can also lead to metastability, where the output remains in an undefined state for an extended period of time.

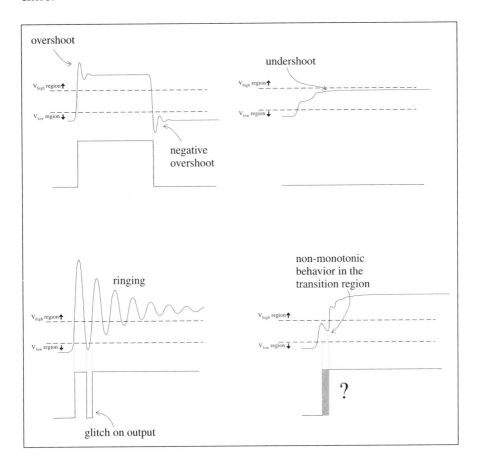

extend further than necessary. Such fields are sometimes called *stray fields*. Transmission line ringing can also exacerbate the crosstalk/radiation problem. As I emphasized in Chapter 7, transmission line effects become apparent in digital systems when the rise time is close to or less than the time it takes for the signal to travel down the transmission line. Limiting transmission line effects not only helps to produce proper

Figure 12.2 Some digital signals and their frequency spectra. The frequency spikes occur at the clock frequency and odd harmonics of the clock frequency. The rise time determines how fast the harmonics decay, and thus determines the overall shape of the curve.

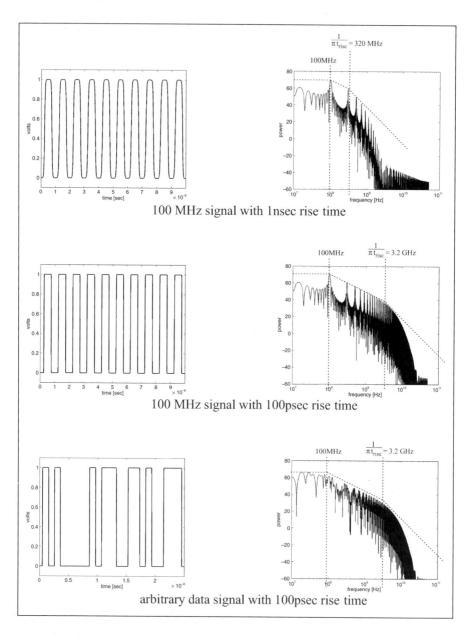

signals, but also limits radiation effects caused by the high-frequency ringing signal.

CONDUCTED VERSUS INDUCED VERSUS RADIATED INTERFERENCE

In EMC literature and EMC regulations the mechanisms of electromagnetic interference are divided into two categories, conducted and radiated. Historically, this categorization has been used in the regulations and, unfortunately, it continues today. I use the word "unfortunately" because both near-field coupling and far-field radiation are lumped under the term *radiated emissions*. While experts in electromagnetics are quite familiar with the difference between near-field coupling and far-field emissions, the vague terminology in EMC regulations tends to confuse those who are trying to learn the basic concepts. As I explained in Chapter 5, near-field energy is stored, not radiated. Furthermore, when you couple to the near-field of a circuit, you not only pick up unwanted electromagnetic energy, you also alter the operation of that circuit by extracting and/or redirecting energy that was otherwise being stored in the neighboring air. The Yagi-Uda antenna, as discussed in Chapter 11, exemplifies this phenomenon; it is constructed by placing two unconnected wires in the near field of a dipole antenna. This confusing EMC emissions terminology leads people to say ridiculous things like, "preventing 60 Hz radiation." Radiation at 60 Hz is never encountered on Earth because you must be over a thousand miles from the source to even get to the edge of the near field. Interference at 60 Hz is caused by near-field coupling. In general, large stray fields can often be eliminated by using uniform transmission lines and by avoiding transmission line discontinuities.

I use the terms *induced interference* for near-field coupling and *radiated interference* for far-field radiation. This distinction is not purely academic. There are many practical differences between how induced and radiated interference occurs. Recall how electromagnetic shields behave in different types of fields. In the near field, electric fields are reflected by a thin metallic shield quite well, whereas magnetic fields readily penetrate metallic shields unless the shield is several skin depths thick. The far-field behavior of shields is different from both magnetic and electric near-field behavior.

Similar effects occur in how circuits pick up near-field electromagnetic energy. Whereas radiated waves always maintain the impedance of air and are therefore always electromagnetic, near-field waves are usually dominated by one component, electric or magnetic. Circuits will pick up radiated energy if they contain antenna-like elements: loops or dangling monopole or dipole antennas. The most energy will be absorbed when the circuit impedance is matched to the self-impedance (feeding impedance) of the antenna. Induced energy coupling has different characteristics. *High-impedance circuits are very susceptible to interference from electric near fields, and low-impedance circuits are very susceptible to interference from magnetic near fields.* Furthermore, circuits with unconnected metal conductors of large size are more susceptible to electric fields, and circuits with large loops are more susceptible to magnetic fields. Both types of induced interference increase when the two circuits are brought closer together.

From these facts, you can conclude that if you want to create a magnetic field meter, you should use a loop or multiturn coil of wire, connected to a current amplifying circuit that has very low impedance. Coupling to a magnetic field can be thought of as coupling to a series voltage source, with series source inductance. The source inductance is the mutual inductance of the two circuits. If you want to create an electric field meter, you should use a metal plate or dangling wire that is connected to a high-impedance voltage amplifier. Coupling to an electric field can be thought of as coupling to a shunt current source, with a parallel source capacitance. The source capacitance is the mutual capacitance. Near fields that are equal in electric and magnetic fields can be thought of as a superposition of electric and magnetic fields; that is, they do not give rise to any new behavior.

The induced interference is also dependent on frequency. The energy in the near field is, perhaps surprisingly, mostly independent of frequency. However, the effects induced into a second circuit are proportional to how fast the field changes (i.e., proportional to the time derivative of the field). High-frequency signals and signals with rapid rise times do not create larger energy in the near field, but they do cause larger induced currents and voltages in any nearby circuits. To summarize, near-field interference increases with larger fields, higher frequencies, and shorter distances.

The equivalent circuit for capacitive coupling (induced interference) and the resulting induced circuit voltage are shown in Figure 12.3. The analogous results for magnetic coupling are shown in Figure 12.4.

The third type of interference is *conducted interference*. Conducted interference consists of unintended signal energy that leaves a product through its cables. For example, high-frequency energy can couple to the power supply and escape the product by traveling out on the power cord. From that point the interference energy can couple to other prod-

Figure 12.3 Electric field (capacitive) coupling and equivalent circuit.

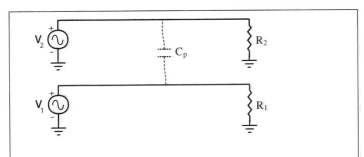

Capacitive coupling between two circuits.

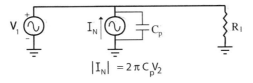

$|I_N| = 2\pi C_p V_2$

Equivalent circuit (first order approximation) from perspective of circuit 1.

ucts directly through the power line. The high-frequency signals that have escaped a product via cabling may also radiate quite readily from the cable. Conducted emissions are controlled mainly by filtering and proper grounding practices.

CROSSTALK

Crosstalk is the unintended induced (near field) coupling of two circuits or transmission lines. In contrast to interference, the term crosstalk is typically reserved for such coupling that takes place within the same product. It is probably fairly obvious that components containing coils (inductors, transformers, motors) are predominated by inductance and therefore give rise to a predominantly magnetic near field. Similarly, capacitors are predominated by capacitance and give rise to a predominantly electric near field. What about cables and PCB connections (i.e.,

Figure 12.4 Magnetic field (inductive) coupling and equivalent circuit.

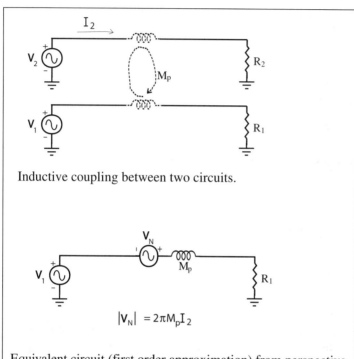

Inductive coupling between two circuits.

$|V_N| = 2\pi M_p I_2$

Equivalent circuit (first order approximation) from perspective of circuit 1.

transmission lines)? The determining equation for the character of the near field surrounding transmission lines is surprisingly simple:

$$\alpha = R/Z_o$$

where R is the terminating resistance on the line and Z_o is the characteristic impedance of the line. If α is small ($\alpha \ll 1$), the near field is predominantly magnetic. If α is large ($\alpha \gg 1$), the near field is predominantly electric. If α is an intermediate value ($\alpha \sim 1$), the near field is electric and magnetic. Because transmission lines are two-port devices, there is a terminating resistance at both ends, R_s and R_L. Therefore, α must be computed at both ends of the transmission line. If α is very different at the two ends, then the near field character will transition from one end of the line to the other. Usually it is the load end that is of concern, but source end–induced crosstalk can propagate to the load as well.

Consider the case of two signals on a ribbon cable or two parallel microstrip traces on a PCB. Because of capacitive and inductive coupling that occurs between the lines, a signal on one line can couple to the other line even though there is no metal (galvanic) connection between the transmission lines. The end result is that part of the signal leaks into another circuit where it acts as noise and can cause errors. Crosstalk can occur via electric or magnetic fields or via a superposition of both. You can determine the character of the coupling by multiplying the near-field coefficients of the two transmission lines, $\alpha_{crosstalk} = \alpha_1 \times \alpha_2$. If $\alpha_{crosstalk}$ is less than 1, the crosstalk takes place through magnetic (inductive) coupling. If $\alpha_{crosstalk}$ is greater than 1, the crosstalk takes place through electric (capacitive) coupling.

PART II: PCB TECHNIQUES

CIRCUIT LAYOUT

A *schematic diagram* shows components and their interconnections in a logical sense. In other words, a component may be placed to the left of another component on the schematic, but the physical placement of the components may be completely different. The companion to the schematic diagram is the layout diagram. A *layout diagram* is the blueprint of the physical product, showing the actual placement of components and the physical connections between them. Modern electronic circuits are constructed on printed circuit boards (PCBs). PCBs are constructed from layers of dielectric with copper traces for components and connections. The simplest PCB is a single layer of dielectric with copper traces on top and bottom. Components are placed on the top and can also be placed on the bottom of the board. Layers of dielectrics can be stacked and pressed into a sandwich. These inner layers are used as additional space for signal routing. Components, of course, can only be placed on the outer (top and bottom) layers. It is not uncommon for PCBs to comprise ten or more signals layers. Connections between signals on different layers are made with vias, which are simply metal-plated holes. Any high-speed or RF PCB should also include layers reserved for ground and power. These ground and power planes are simply sheets of copper that span the entire area of the circuit board. The use of such planes allows transmission lines to be created on the

PCB. Sometimes the outer layers are used solely for components, with the remaining area being covered with grounded copper. In such boards, all signaling occurs on the inner layers, and the outer grounding layers serve as a shield for the board.

PCB TRANSMISSION LINES

Proper use of transmission lines is the key technique to reducing unwanted parasitics, radiated emissions, emission susceptibility, and crosstalk. *In other words, proper use of transmission lines is the most important design tool for maintaining EMC and signal integrity.* Electronics designers should not view EMC as something that can be handled with shields, ferrites, and so on. Although these techniques are useful, they work more as a remedy than as a cure. Because of the importance of transmission lines, it is paramount for electronics designers, EMC engineers, and layout engineers/technicians to have a good understanding of PCB transmission lines. Figure 12.5 shows the common PCB transmission lines. Microstrip is one of the most common, so I will focus my discussion on microstrip. The other types of PCB transmission lines exhibit very similar behavior to microstrip.

The exact equations for characteristic impedance and wave velocity of microstrip transmission lines are very complex because the wave

Figure 12.5 PCB transmission lines.

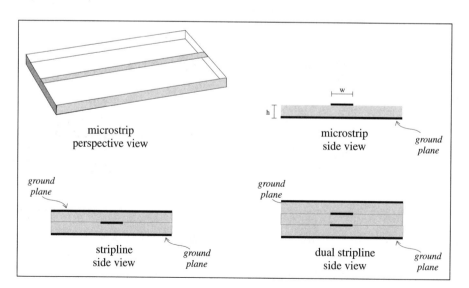

energy travels partly in the dielectric of the PCB and partly in the air. I will therefore cover the formulas in only a qualitative manner. For the formulas, refer to Pozar (1998), Johnson (1993), or Paul (1992). The characteristic impedance of microstrip lines is proportional to the thickness of the dielectric between the signal trace and the ground plane. This parameter is called the height, h, of the trace above the ground plane. The characteristic impedance is inversely proportional to the width of the copper trace, and is inversely proportional to the square root of the relative dielectric constant. These three relations can be combined into a single formula as

$$Z \propto \frac{h}{w\sqrt{\varepsilon_r}}$$

An important characteristic of this formula is that the impedance scales with size. In other words, if you have a 50 ohm line, and you shrink the width by 1/2, you then can reduce the height by 1/2 to produce a smaller line that still has an impedance of 50 ohms. For this reason, RF PCB layers need to be thin, or the 50 ohm traces are unreasonably wide.

There is an even simpler way of understanding how characteristic impedance depends on geometry, using the general relation for characteristic impedance:

$$Z_o = sqrt(L/C)$$

Making a trace wider reduces inductance in the same way that it reduces the resistance of the trace. Furthermore, the microstrip trace and ground plane together form, in effect, a plate capacitor, and by increasing the width of the trace, you increase the effective area of the capacitor. Because $C = \varepsilon A/d$, the capacitance is increased. Using the general relation for Z_o, you can see that both effects cause a reduction in the characteristic impedance.

In contrast, if you increase the distance between the trace and the ground plane, the capacitance goes down, and the inductance goes up. Therefore, increasing the height causes an increase in the characteristic impedance.

Wave velocity is a function of the dielectric constant of the material the energy travels through. For microstrip, the energy is split between the PCB dielectric and the air, and thus the velocity takes a value intermediate to the values in the dielectric and air. The most common dielectric for PCBs is called FR4, a fiberglass-resin laminate (the FR stands for flame retardant). The dielectric constant (ε_r) of FR4 is between 4.0 and

4.6 depending on the exact manufacturing materials and process. As a consequence, the velocity of microstrip traces on FR4 is between 47% and 100% of the speed of light, c. For traces on inner layers, the velocity is exactly

$$v = \frac{c}{\sqrt{\varepsilon_r}}$$

Recall, from Chapter 1, the fundamental relation between velocity and wavelength, $\lambda = v/f$. Therefore, *the wavelength of a signal on a PCB trace is less than that in air.* This fact must be taken into account when determining electrical length of transmission lines. Signal delay along PCB transmission lines is also affected by the velocity. The delay, τ_{delay}, can be expressed in two equivalent ways,

$$\tau = \frac{\ell}{v} = \frac{\ell_\lambda}{f}$$

where ℓ is the length and ℓ_λ is the electrical length of the transmission line.

Eagleware (www.eagleware.com) sells a very useful computer application, TLINE, that can find the necessary trace width for a given characteristic impedance. It also calculates important parameters such as wavelength, velocity, and maximum useable frequency.

THE PATH OF LEAST IMPEDANCE

One of the most fundamental relations taught in basic circuit theory is that current follows the path of least resistance. At high frequencies this relation is modified to "current follows the path of least impedance." *The path of least resistance and the path of least impedance may or may not be the same, depending on the circuit geometry and the signal frequency.* In any circuit, the signal consists of the flow of two currents, typically called the signal and return. Since the return current often flows through ground, designers often neglect to think about the return current. This is a bad habit to fall into when in high-frequency design. The signal current and return current form a loop, and therefore produce an inductance. The inductance is proportional to the area of this loop. Inductance produces an impedance that increases with frequency, $Z = 2\pi f L$. Therefore, large loops impede high-frequency current flow. At high frequency, the return current flows most easily when the return path

Figure 12.6 Microstrip circuits demonstrating the concept of path of least impedance. In each case the high-frequency current takes a different path from the low-frequency current.

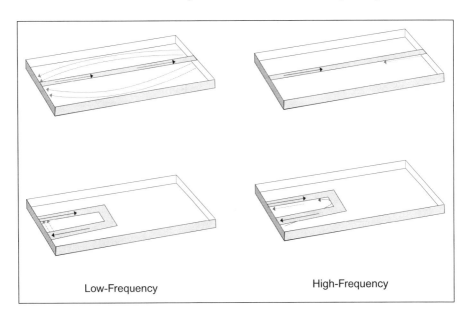

follows the signal path as closely as possible. Figure 12.6 exemplifies this concept—the high-frequency return current on microstrip transmission lines travels directly underneath the trace, in contrast to the low-frequency current which spreads out along the ground plane.

The simple solution to the problem of high-frequency design is the use of uniform transmission lines. However, because of connections to components, cable connectors, layout constraints, and other reasons, perfectly uniform transmission lines are not possible. For example, when a signal connects from a microstrip transmission line to a surface mount integrated circuit (IC), the return current must somehow travel from the surface of the PCB down through a via and onto the ground plane. From that point, the return current will take the shortest path to where it can flow underneath the signal trace. If the signal trace is not on the surface of the PCB, then it too must follow a via to another layer. Figure 12.7 illustrates this problem.

Strange and unwanted effects can happen if current loops exist in the path of high-frequency signals. First of all, the loop will appear as an inductance to the driving amplifier (analog circuit) or the driving logic gate (digital circuit). This inductance can cause analog amplifiers to

Figure 12.7 Examples of signal traces that jump across ground layers through a via. The return current is designated by the dotted arrows.

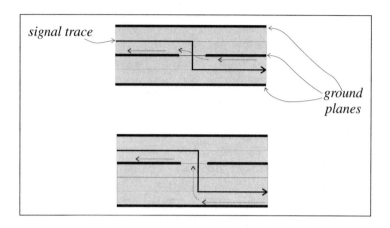

oscillate and can cause ringing and overshoot on digital signals. The second problem is that of crosstalk. The loop produces a large stray magnetic near field that is associated with the inductance. This near field can couple the signal to nearby circuits. The third problem is that of radiation. The loop acts like a small loop antenna, and radiates energy readily. *Unintentional circuit loops are one of the leading causes for products to fail EMC emissions testing.*

When confronted with inductance, a signal may also take a parasitic capacitive route as an alternative. Keep in mind that the impedance presented by a capacitor decreases with frequency. This fact leads to some of the strange behavior of high-frequency signals. Unlike DC signals, which must have a conducting circuit to propagate, high-frequency signals often have an easier time propagating through air; high-frequency signals have a tendency to couple capacitively through air and to radiate through air. High-frequency signals will only follow conductors if uniform transmission lines or waveguides are used.

THE FUNDAMENTAL RULE OF LAYOUT

The fundamental rule of layout is to *know the path of the return current*. When dealing with PCB layout and component connections, identifying the return path takes some practice. The practice pays off, however, and eventually identifying the return path becomes second nature. For low-frequency, high-power signals, you must make sure that the return path has a low resistance. For high-frequency signals (both high-speed digital and RF), you must make sure the return path has low inductance—you must minimize loop area.

At high frequency, current will always flow on the ground plane directly underneath the signal trace, if such a path is available. Therefore, never route traces over cuts in the ground plane. Furthermore, make ground connections to components such that the return signal has the shortest path to make its way underneath the signal traces. This guideline implies that vias to the ground plane should be as close as possible to the IC ground pins. You should also consider using clusters of two or more vias when making ground connections so as to reduce via inductance and to provide the shortest return path for signals on different sides of the IC.

SHIELDING ON PCBS

Good layout practice is sufficient for the vast majority of circuits, but certain situations do require the use of shielding. PCB shielding techniques can be divided into two categories, true shielding techniques and diversion techniques. True shielding techniques simply involve enclosing the circuit inside a metal box. Power supplies (AC-DC, DC-DC power converters, etc.) are good examples. PCB mountable power supplies are often sold with their own metal enclosure to provide shielding. Although power supplies rarely involve high frequencies, they can produce large near fields and large radiation. There are two major reasons for this. First, power supplies usually manipulate large currents, which can cause large fields from their magnitude alone. Second, although DC-DC converters typically operate at frequencies on the order of 100 kHz, they involve square wave signals that may have rapid edge rates. The harmonics associated with the edge rates can exceed 50 times that of the fundamental frequency of operation. Therefore a DC-DC converter operating at 500 kHz, may actually produce fields and radiation up to 25 MHz or higher.

In addition to components purchased with integrated shields, circuits on PCBs can also be shielded. PCB shields can be purchased that consist of a metal box with five sides. The sixth, open, side is placed over the circuit and is attached to the PCB. The ground plane of the PCB then forms the sixth side of the shield. This technique is often used for the RF front-end circuitry of TVs and VCRs. It is very important to make a good connection at multiple locations between the PCB and the metal shield when using this technique. Otherwise, the metal shield can cause different signals in the circuitry to capacitively couple via the shield, as explained in Chapter 9.

Signals can be shielded by placing the signal traces on internal layers, with grounds on the outer (top and bottom) layers. These form stripline transmission lines. Such techniques are often recommended for signals above 100 MHz. To reduce radiation effects from PCB edges, all inner-layer nongrounded metal (both signal layers and power plane layers, but not ground layers) can be limited to a maximum distance of $20h$ from the PCB edge, where h is the distance between the power and ground planes. The outer layers are then covered in ground plane all the way to the edge of the board, forming an efficient shield.

Diversion shielding is accomplished by placing traces of grounded copper around the circuit. Technically this type of design doesn't perform as a shield. Instead, it serves to concentrate and divert the electric field lines of the circuit it surrounds. The same technique can be used for magnetic fields by using high-permeability materials, such as mu-metal. High-permeability material shields are available on cables and as enclosures, but not for PCBs.

Applying this concept in RF circuits, grounded metal is sometimes used as fill between the signal traces. The grounded areas should be connected to the ground plane with vias ("stitched") such that there are many vias per square wavelength. This technique changes the transmission lines from microstrip to microstrip with coplanar ground, also known as grounded coplanar waveguide. The characteristic impedance of the transmission line is lowered by using this technique because the guard traces increase the capacitance between the signal trace and ground. Figure 12.8 shows how the field lines are changed. This technique can be extended to the handling of RF ICs. By placing a ground underneath the entire area of an IC, the input and output are prevented from capacitively coupling to one another. This technique is very useful for RF switches and active filters, where isolating the input and output is very important.

Grounded traces are sometimes used to surround digital clock signals, which tend to have fast edge rates. In addition to having a ground plane below the signal trace, copper traces are placed on either side of the trace and are stitched to the ground plane in multiple locations. In digital systems this technique is called guard traces. For digital systems this technique is really useful only for traces on outer layers. On inner layers it makes more sense to just leave some empty space around the trace (3 W rule), as discussed later in this chapter. Because digital PCBs typi-

Figure 12.8 Grounded coplanar waveguide (microstrip with coplanar ground).

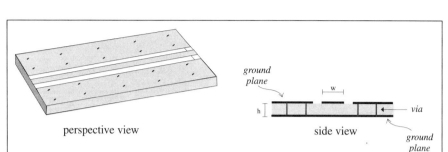

cally have many more signal connections than RF PCBs, it is not feasible to ground all the regions between the traces.

Finally, any unused areas of PCB (places where there are no signal traces) should be filled with grounded metal. This serves to concentrate the electric fields from the traces, and it shields the signals of the inner layers of the board.

COMMON IMPEDANCE: GROUND RISE AND GROUND BOUNCE

Consider the circuit shown in Figure 12.9. Two circuits are connected to the same ground wire. Since all grounds are made from non-ideal wire, they will have some resistance. The circuit on the right is a high-power, high-current circuit. Its return current flows back to the ground wire to the power supply. The return current is so large that it causes a considerable voltage drop across the resistance of the ground wire. The end result is that the circuit on the left is referenced to a voltage that is above the ground potential. The input and output of the left circuit presumably connect to other circuits. Assuming that the other circuits refer to the actual ground level of the power supply, an error will be produced. This effect is called *common impedance coupling*. If the high-current circuit varies in time, the inductance of the ground wire will also cause changes in the ground. A large sinusoidal current, like that of an RF transmitter, will cause the ground to be modulated at the frequency of the oscillation.

In digital circuits the effect is called *ground bounce*, because it typically occurs when a signal changes from high to low level (or vice versa). The

Figure 12.9 Demonstration of common impedance coupling, ground bounce, and star grounding.

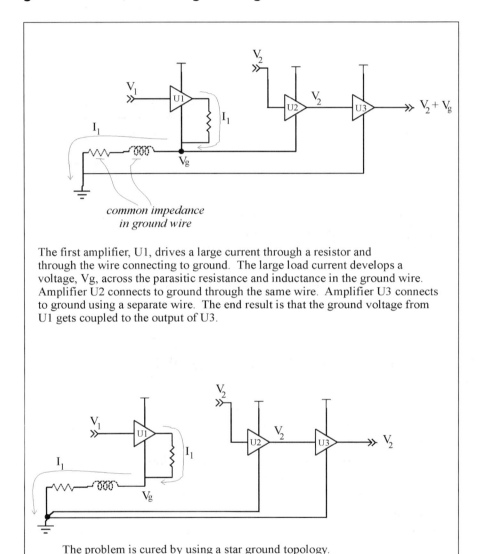

The first amplifier, U1, drives a large current through a resistor and through the wire connecting to ground. The large load current develops a voltage, Vg, across the parasitic resistance and inductance in the ground wire. Amplifier U2 connects to ground through the same wire. Amplifier U3 connects to ground using a separate wire. The end result is that the ground voltage from U1 gets coupled to the output of U3.

The problem is cured by using a star ground topology.

edge of a digital signal has considerable high-frequency content. The high-frequency return current within the edges can cause noticeable ground voltage changes due to the inductance of the ground wire. Once the edge current has returned, the ground voltage goes back to its steady-state value—hence the term ground bounce. Ground bounce can easily cause glitches to occur. Moreover, it may not be a single signal that causes the ground bounce. If an entire group of signals, such as a 32-bit bus, all change value at once, the ground surge may be tremendous. This problem is referred to as *simultaneous switching noise*. Such problems are intermittent and difficult to track down. It is therefore imperative to have good ground and power supply design performed during layout of the PCB.

STAR GROUNDS FOR LOW FREQUENCY

The best way to avoid ground problems at low frequency is the use of a star or single-point ground. This technique involves using a separate wire for the ground connection of each IC. The ground wires are tied together at a single point, typically at the output of the power supply. Since none of the circuits share a ground wire, common impedance coupling is eliminated. This technique works well for audio frequencies (<20 kHz) and can also perform well up to about 1 MHz.

DISTRIBUTED GROUNDS FOR HIGH FREQUENCY: THE 5/5 RULE

At higher frequencies, the inductance caused by the star ground wires creates more problems than it solves. The best ground solution for high-frequency designs, both RF and digital, is the *distributed ground* (also called a multipoint ground). The most common example of a distributed ground is the use of an entire sheet of metal—that is the ground plane. The main benefit of the ground plane is that it allows uniform transmission lines to be used throughout the design, which reduces radiation and inductive loops. The ground plane also prevents ground bounce and other grounding problems; there is such a large amount of conductor that the ground resistance and ground inductance are extremely small. Ground planes can really be used in any design, not just high-frequency designs. The major trade-off in using ground planes in low-frequency design is cost. Every extra layer on a PCB increases cost, so ground planes are not usually used in consumer audio equipment.

Another useful option for boards that don't use a ground plane is the use of power and ground buses. In this design the power and ground are carried on wide parallel traces that run along the entire design. Each IC is connected to the power and ground with very short traces. Grounding problems are prevented by making the ground trace very wide. This technique is also commonly used for the circuitry inside ICs. An even better option is to create a power/ground grid with, for example, vertical traces on the top layer for power and horizontal traces on the bottom layer for ground. A decoupling capacitor is placed at each intersection to provide low impedance between the planes.

The 5/5 rule, as advocated by Kimmel and Gerke (1997), can be used for deciding whether or not a ground plane is necessary for your design. This guideline states that if any signal has a frequency greater than 5 MHz or has edges faster than 5 nsec, you should use a ground plane.

TREE OR HYBRID GROUNDS

In many products and systems, both types of grounding techniques (star grounds and distributed grounds) are used. Such grounding systems are referred to as hybrid grounds; however, I prefer the term "tree ground" because I think it is more illustrative. The tree ground starts at the main power supply for the product and then branches to different subsystems. Each subsystem then typically uses a local distributed ground. The branching provides isolation between different parts of the design. High-current circuitry such as heater circuits, RF transmitters, relays, and motors should have their own branch so that these high currents do not disrupt other, more sensitive circuits. It is often important to separate digital grounds from sensitive analog circuits. RF circuits should be separated from low-frequency data or audio. The chassis or safety ground should also have its own connection. The safety ground connection should be made as direct as possible to the ground supplied by the power cord. There are two reasons for this. The first reason is that of safety. By making direct connection, safety is not compromised if any other ground connections are broken. The second reason is that of noise immunity. The product chassis serves as shield, and as such it may carry noise currents such as those induced by electrostatic discharge (ESD), or RF currents picked up by cables. You certainly want to avoid coupling such noise into any circuitry.

Each of these ground branches can be and usually is implemented using ground planes, but the different ground planes are only connected at one place (or very few places). Interface signals between the different subsystems must be handled with care. On a PCB, these interface signals can be routed with the "moat and drawbridge" technique. The isolation (moat) of the ground planes is broken by a copper connection (drawbridge). All interface signals should travel over this drawbridge so that each signal forms its own transmission line with grounded drawbridge. Never route a signal over a moat without a drawbridge!

A word of caution—isolation of grounds on PCBs should only be used when it is certain to be needed. Overuse of ground islands, drawbridges, and so on can lead to nightmare designs.

POWER SUPPLY DECOUPLING: PROBLEMS AND TECHNIQUES

Power supply decoupling is the term given to the technique of making sure the DC power line variations do not affect the loads (amplifier, ICs, logic gates, etc.) and vice versa. Since most ICs have AC signals as inputs and outputs, the current drawn from the power supply will vary in an AC manner. For example, an RF circuit that amplifies a 900 MHz signal will draw a supply current that varies at 900 MHz. A digital CMOS circuit that buffers a 100 MHz digital signal will also draw a supply current that varies at 100 MHz. For digital CMOS circuits, the vast majority of the current occurs during the transitions between the high and low voltages. Therefore the edges of digital signals can create very short time periods of high current demand. Power supply circuits themselves can only handle relatively low-frequency current variations. A typical power supply can track variations of from several hundred to several thousand Hertz. Furthermore, due to the physical distance between the supply and the circuit load, varying currents can propagate along the supply lines. The varying currents cause the voltage of the supply to sag and surge. If the voltage of the supply varies too greatly, the IC that is using the supply will not function properly. A second result is that the supply voltage becomes modulated. The supply modulations travel to other ICs, and signals then couple from one IC to another through the supply. To address these problems, the supply and loads are *decoupled* from each other by placing capacitors across the supply and ground.

Power supply decoupling has two related goals: charge supply and filtering. The charge stored on the decoupling capacitors serves as a reservoir that is quickly accessible for load variations. Each circuit has an effective impedance that it presents to the power supply; that is, each circuit acts like a load to the power supply. This impedance is not constant, but depends on the signals that the circuit processes. The circuit will have a DC (quiescent) supply impedance and a variable supply impedance. When a power distribution system is properly designed, the

Figure 12.10 Power supply filtering.

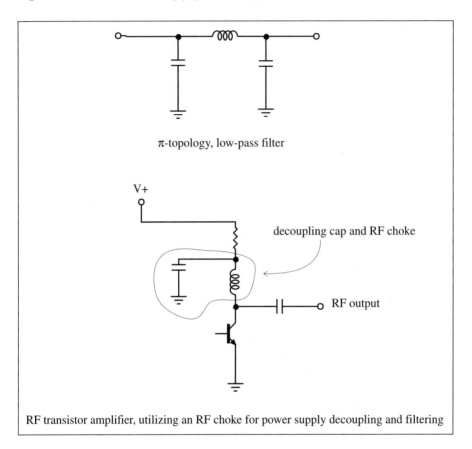

RF transistor amplifier, utilizing an RF choke for power supply decoupling and filtering

decoupling capacitors provide charge for the variations in the circuit loads such that each circuit receives a constant supply voltage at all times.

The second goal or function of power supply decoupling is filtering. All switch-mode power supplies are inherently noisy. Furthermore, any product powered form the 60 Hz mains voltage will receive noise conducted through the power line. Decoupling capacitors serve to shunt this noise to ground. Voltages from signal traces and from ICs can also couple to the power supply through conductive pathways or by near-field coupling. Decoupling serves to shunt these voltages to ground, maintaining a constant DC voltage on the supply. Some circuits require more than just capacitors to provide proper filtering. Very sensitive circuits such as analog-to-digital converters (ADCs) and digital-to-analog converters (DACs) require LC filters to filter the power supply. The best filter topology for this application is that of a pi low-pass filter, as shown in Figure 12.10.

A ferrite bead can be used in place of or in conjunction with the inductor in this circuit. Often the first capacitor in the filter is not explicitly used, but is provided by other decoupling capacitors on the power supply. However, never omit the second capacitor (the load side capacitor), and don't use a T low-pass filter, because such filters will not provide a proper charge reservoir to the load. (The source side capacitor's charge is impeded by the inductor.) LC filters can also be used at the power supply output and at the power entry point to the product. A similar filter arrangement is used in most RF amplifiers. In these applications, the inductor is referred to as an RF choke. The circuit is essentially the same. Clocks and phase lock loops (PLL) also typically require an LC filter to prevent the large oscillations from coupling to the power supply. In low-frequency analog circuits, such as op-amp circuits, the power supply is sometimes filtered with an RC low-pass filter. The resistor lowers the voltage the op-amp receives, so its value must be small (10Ω to 100Ω) to avoid a large voltage drop.

To maximize its utility as a charge storage device, you want to maximize the charge-to-volt ratio of the capacitor, $Q/V = C$. To maximize its performance as a filtering element, you want to minimize the impedance, $Z = 1/jwC$. Both of these goals are attained by maximizing the capacitance, C. Using huge capacitors seems to be a perfect solution. Unfortunately, decoupling is not that simple. Part of the problem stems from the parasitic elements of capacitors, as discussed in Chapter 7. Any real capacitor has a series resistance (ESR) and series inductance (ESL). ESL is mostly a function of package size. Large capacitors need large packages and thus larger inductance results. ESR is mostly a function of the dielectric material. Large capacitors are typically electrolytic, such as aluminum electrolytic and tantalum electrolytic, because they can produce large capacitance in a smaller physical size. Unfortunately, electrolytic capacitors have poor tolerance, are unipolar (voltage can be applied in only one direction), have poor temperature stability, and have larger internal losses. This last fact equates to large ESRs. Ceramic materials are used for high-frequency capacitors because they perform better in all areas, except that their dielectric constant is lower.

Referring back to Figure 7.4, the frequency response of a capacitor has two distinct regions, which are separated by the resonant frequency. In the low-frequency region, a capacitor behaves like a capacitor, as intended. Its impedance decreases in inverse proportion to frequency. At resonance, the capacitance and the ESL exactly cancel one another, and the impedance is equal to the ESR. In the high-frequency region, a capacitor behaves as an inductor, thus its impedance increases with frequency. As the impedance rises, the capacitor becomes less and less useful as a decoupling device. The inductive reactance of the capacitor at high frequencies causes another, more problematic, effect—antiresonance. Antiresonance occurs when two or more capacitors with different resonant frequencies are placed in parallel. Consider the case of two capacitors, C_1 and C_2, that have different capacitances but equal ESL. Capacitor C_1, the larger capacitor, has a resonance at frequency, f_1, above which it behaves like an inductor. Capacitor C_2 has a resonance at a higher frequency, f_2, above which it behaves like an inductor. Between these two frequencies, C_1 acts like an inductor and C_2 acts like a capacitor. At some point in between the two resonant frequencies, antiresonance occurs. Antiresonance occurs at the parallel resonance of the inductance of C_1 and the capacitance of C_2. Near this frequency, the impedance rises dramatically, hampering the decoupling ability of the two capacitors. The effect is shown in Figure 12.11.

There are two ways to decrease the height of the anitresonance. The first method is to reduce the frequency gap between the two resonant frequencies. The second method is to use capacitors with higher ESRs. This second method has the side-effect of raising the impedance of the resonance. A balance between the two problems must be found for optimal performance.

The interaction of several capacitors on an actual PCB at high frequencies is not as easy to calculate because the PCB adds its own effects. The traces and vias that connect the decoupling capacitors to the power and ground planes have impedance that is mostly inductive. This inductance further limits the frequency response of the capacitors. Furthermore, the current paths along the power/ground planes themselves cause transmission line effects at high frequency when the distances traversed by the current are electrically long.

To properly model decoupling at high frequencies, the PCB must be included. Therefore, some of the conclusions from studies on decoupling that model only the capacitors themselves cannot be trusted. At low frequencies (when the dimensions of the power/ground planes are less than $\lambda/20$), the planes can be modeled as contributing about 0.1 nH per inch that a decoupling current travels on the plane. When the planes become electrically large in dimensions, proper modeling takes the form of modeling the power and ground planes as a two-dimensional transmission line. A two-dimensional transmission line has length and width. Signals propagate from their source in two dimensions, as opposed to commonly used transmission lines such as

Figure 12.11 When capacitors of different value are placed in parallel, antiresonant peaks occur. Using two capacitors of the same value produces better results.

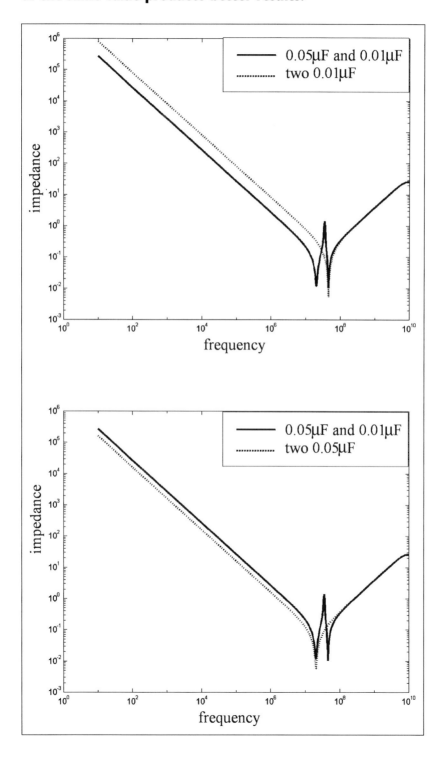

wire pairs that transmit in only one dimension. The capacitor models should include the ESL and ESR. The vias and traces that connect the capacitors to the power/ground planes can be modeled as inductors. The loads can be simulated by using AC or transient current sources that inject current at discrete locations on the PCB. Then the resulting voltage at all points on the plane can be computed to determine the effectiveness of decoupling. If the resulting voltage is lower than the maximum ripple allowed by the components, your design is a success. *Spice* is often used as the circuit simulation product once the models are created. Such modeling is definitely not required for all PCB products, but such extensive modeling must be used for any high-frequency decoupling studies to have merit. Refer to the papers by I. Novak in the bibliography for more information on accurate power distribution modeling.

Traditionally, decoupling design for digital systems was quite simple. You placed a 0.1 μF capacitor at each IC, and then included a large bulk capacitance (from 10 μF to several thousand μF) somewhere on the board near the power supply. Unfortunately, such simple rules of thumb are not sufficient in digital designs in the hundreds of MHz and higher. Such designs typically have signal rise times less than 1 nsec. Even designs with clock frequencies as low as 30 MHz can cause problems when using the old rules of thumb. Other changes have also affected decoupling. The leading edge microprocessors can have tremendous current surges (10 A or more). In addition, the voltages for digital systems have been successively lowered from 5 V to 3.3 V to 1.8 V and lower. The unfortunate result of lower supply voltage is that the tolerable supply voltage ripple has also been lowered. Whereas in the old days of 5 V TTL logic ICs could handle 500 mV of ripple, some of the latest low-voltage technologies can only handle about 50 mV of ripple. Thus, there are three reasons why decoupling is becoming more difficult: higher speeds, larger currents, and smaller ripple voltage tolerance.

With all these problems, there are some new technologies that result in better decoupling. The first technology is that of organic semiconductor (OS-CON) electrolytic capacitors. These capacitors combine the high-density capacitance of electrolytic technology with low ESR. Sanyo makes a line of these capacitors. The OS-CON technology produces capacitors with up to thousands of μF that have inductances from 1 nH to 5 nH and ESRs down to about 0.01 ohms. Compared with previous technologies, the ESL is about one quarter that of traditional electrolytic caps and the ESR is about one hundredth that of traditional electrolytic caps. These caps can provide bulk decoupling with much lower impedance.

POWER SUPPLY DECOUPLING: PROBLEMS AND TECHNIQUES

Advancements have also been made in the technology of capacitors for high frequency. The materials of choice for high-frequency applications are ceramics, such as NPO (a.k.a., C0G) and X7R. The problem for high-frequency capacitors is that of parasitic inductance. New technologies are all aimed at reducing the ESL of high-frequency capacitors. Typical ceramic SMD capacitors are about twice as long as they are wide. SMD capacitors in the size ranges of 0603 to 1210 typically have ESLs in the range of 0.750nH to 1.250nH. A simple technical advance for surface mount capacitors is that of swapping length and width. For example, instead of manufacturing capacitors that are 80 mils long and 50 mils wide (0805), companies are now manufacturing capacitors that are 50 mils long and 80 mils wide (0508). The wider package gives a wider lead with lower inductance (about 40% lower). Another technology is that of interdigitated capacitors (IDC) and low inductance capacitor arrays (LICA). This technology uses parallel capacitors in an integrated package. The capacitors are arranged such that current flows in opposite directions in adjacent capacitors. The magnetic fields of adjacent capacitors thus tend to cancel each other, resulting in very low inductance, under 1 nH and as low as 50 pH. IDC capacitors use standard SMD mounting, but the LICA caps use leads on the bottom of the package, basically a ball grid array (BGA), to further reduce inductance. AVX (www.avxcorp.com) manufactures such capacitors and provides extensive application notes on these technologies.

The technology of the PCB has also been improved. As a rule, separate power and ground planes are used on modern digital PCBs (and are probably a necessity for systems with clock frequencies above 5MHz). If placed adjacent to each other in the PCB layer stackup, the two planes can provide charge storage and thus decoupling. Two planes separated by 10 mils of FR4 can provide about 50 pF per square inch of coupling capacitance, with inductance mostly limited by the vias to the IC leads. Using smaller spacings and higher dielectric constants, capacitance as high as $2.5\,nF/inch^2$ can be attained. These techniques are referred to by several names: power plane cores, buried capacitance, or integrated capacitance. Hadco and its subsidiary Zycon are two of the companies that support these technologies.

Because the power/ground planes form a 2D transmission line, reflections and radiation can occur at the edges of the board. Reflections on the power/ground planes limit the effectiveness of the planes for decoupling, mostly because of standing waves. To reduce these effects, discrete resistors or buried resistance strips can be used to connect the power and ground planes along the edge of the PCB. This technique is called dissipative edge termination (DET).

IC manufacturers are also learning that as digital speeds get ever higher, the ICs themselves must have decoupling capacitors integrated onto the chip. On-chip decoupling is a necessity for GHz digital ICs.

There are also layout techniques that can improve high-frequency decoupling performance. Placing the decoupling capacitor as close as possible or underneath the IC on the bottom side of the PCB allows for the highest-frequency performance. This technique reduces the path of the current and reduces inductance. The capacitors should be connected to the IC leads using the shortest and widest traces possible or should be connected directly to the ground and power planes using multiple vias. To achieve the best possible performance, you can use both techniques: connect the capacitor to the IC leads using traces and by using multiple vias to and from the power/ground planes. A single via has a typical inductance of 0.2 nH to 0.5 nH. A trace on the surface of a PCB has a typical inductance of 20 nH/inch to 30 nH/inch. By using multiple vias in parallel, and placing the vias as close as possible to the capacitor pads you will get the best results. Some manufacturers even allow vias to be placed inside the pad.

POWER SUPPLY DECOUPLING: THE DESIGN PROCESS

Power supply decoupling can be broken up into two general frequency regimes, low and high. At low frequency, decoupling capacitance is usually referred to as *bulk capacitance*. I refer to the high-frequency decoupling as *local decoupling* because for it to be effective, high-frequency decoupling capacitance must be placed close to the IC being decoupled.

In the low-frequency region, the PCB itself is electrically short; therefore placement of the bulk capacitors is mostly irrelevant to performance. There will be some ohmic (I^2R) losses as the current travels from the bulk capacitors to the loads, but this is typically negligible. As you may recall from Chapter 8, when a transmission line is electrically short, the signal is able to propagate to all points on the transmission line faster than any significant change can occur in the signal. Loosely speaking, in low-frequency decoupling the ripple current can find its way to the decoupling caps in a time scale that is much shorter than any variations in the ripple signal itself. The bulk capacitance will provide approximately equivalent performance whether it is spread evenly about the board or lumped in some specific location. For a PCB with a largest dimension of about 10 inches and signals under 50 MHz, the PCB can be categorized as being electrically short ($<\lambda/10$). For transients and

POWER SUPPLY DECOUPLING: THE DESIGN PROCESS

digital rise times greater than 10 nsec, the PCB can be categorized as being electrically short. From these values, you can determine which capacitors need to be placed near the ICs and which capacitors can be placed arbitrarily. To be a little clearer, a capacitor whose useful frequency range is the range where the PCB is electrically short can be placed arbitrarily. I refer to these capacitors as bulk capacitors. All other capacitors, the local decoupling capacitors, should be placed as close as possible to the source of the power supply variations.

When designing the decoupling systems for a PCB, the first design parameter to determine is the total bulk capacitance needed to handle the current transients. A good estimate to start with can be calculated using the derivative of the capacitance equation,

$$\frac{dQ}{dt} = I = C\frac{dV}{dt}$$

Solving this equation for C yields,

$$C = \frac{I_{transient} \times t_{transient}}{V_{maxripple}}$$

where $I_{transient}$ is the worst-case current transient, $t_{transient}$ is the rise time of the transient, and $V_{maxripple}$ is the maximum allowable ripple. This is the value for a perfect capacitor, without any ESL or ESR.

For an example, consider a microprocessor that can have current transients as high as 10 A, within a 1 μsec time period. The maximum allowable ripple for the microprocessor is 100 mV. Using the previous equation for this example results in

$$C = \frac{10\,A \times 40\,\mu sec}{100\,mV} = 4000\,\mu F$$

At least 4000 μF of capacitance will be needed for bulk decoupling. This approximation will likely be very low because it assumes that the capacitors are ideal. Assume that you chose to use 1500 μF capacitors with ESR = 50 mohms and ESL = 5 nH. Determine the number of these real-world capacitors needed to meet these specifications by using the following equation

$$N \geq \frac{I_{transient}}{V_{maxripple}}\left[\frac{t_{transient}}{C} + \frac{ESL + L_p}{t_{transient}} + ESR\right] = 7.61$$

Figure 12.12 Frequency domain impedance of eight 1500 µF capacitors.

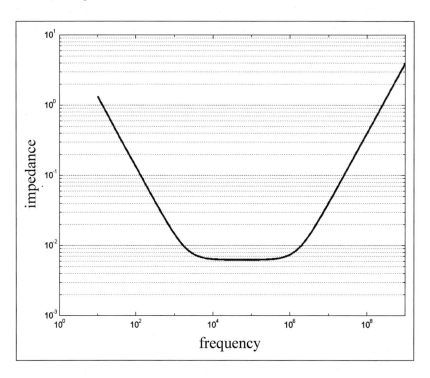

where L_p is the inductance of the connection from the PCB surface to the power/ground plane, and where C is the capacitance, ESL is the inductance, and ESR is the resistance of a single capacitor. Choosing the next highest integer produces $N = 8$, and provides a total of 12,000 µF of capacitance.

The same analysis could also be performed in the frequency domain. First determine the maximum power supply impedance that can be tolerated:

$$Z_{max} = \frac{I_{maxripple}}{V_{transient}} = \frac{0.1V}{10A} = 0.01\Omega$$

The capacitors need to provide an impedance below Z_{max} up to a frequency of about

$$f = \frac{0.5}{t_{transient}}$$

From Figure 12.12, you can see that eight 1500 µF capacitors provide a low enough decoupling impedance, down to about 1.5 kHz, and the high-frequency response stays below Z_{max} until about 1.4 MHz. If the frequency range of the bulk capacitance needs to be extended higher in frequency, smaller value capacitors with lower inductance can be paralleled with the higher value capacitors. A common combination for digital systems is a bank of 1500 µF, 10 µF, and 0.22 µF capacitors, with several capacitors of each value.

When using different values of capacitors, it is very important to plot the results in the frequency domain, including the inductance of the PCB connections. That way antiresonant problems can be identified and addressed, if necessary. Keep in mind that using values smaller than 0.22 µF for bulk capacitors or trying to bulk decouple frequencies much higher than 25 MHz is pointless and may actually cause more harm than good. Local decoupling must be used at high frequencies.

Power Supplies

So far I have not mentioned the specifications of the power supply. Since the bulk capacitors' impedance increases toward the low-frequency end, the lower bound of frequency performance is determined by the specifications of your power supply ripple and response time. For this example, I assumed that the power supply can effectively regulate 100 mV of ripple up to 2 kHz. You may find in practice that the power supply will need decoupling and/or filtering to achieve acceptable ripple. This filter should be placed as close to the supply as possible.

Local Decoupling

For ICs that produce power supply variations less than 25 MHz, large bulk capacitors, greater than 1 µF, are best suited. They can be placed at the IC, but they can also be treated as bulk capacitors because the PCB is electrically short at these low frequencies. Choosing decoupling caps for digital systems depends on the clock frequency, but more importantly, it depends on the edge rates. For digital clock frequencies under 25 MHz, 0.1 µF serves as a good value to use. A value of 0.01 µF serves well up to clock frequencies of about 250 MHz. These are rough guidelines. You should run some Spice simulations to verify performance in your design. For digital systems running above 250 MHz, you must use LICA capacitors or buried capacitance to obtain decoupling. Studies (Greb, 1995) have shown that standard SMD decoupling capacitors less than 0.005 µF have little if any benefit, except in narrow-band RF applications (see the following paragraphs). The problem is that the combination of small capacitance and non-zero inductance limits the low-impedance region of the device to frequencies in the immediate neighborhood of resonance. For digital decoupling above 1 GHz, power/ground plane capacitance and on-chip techniques must be used. For ICs with a single voltage and single ground lead, one cap should usually suffice. For ICs with multiple voltage and ground leads, several decoupling caps are necessary. The best approach is probably the use of multiple capacitors of the same value. When small decoupling capacitors (<0.1 µF) of different values are used in parallel, the antiresonances that result can often cause more harm than good. This topic is one of great controversy. I won't say never parallel local decoupling capacitors of different values, but I do caution against it. Of course, no rule of thumb can compare to calculation/simulation.

RF DECOUPLING

In general, the techniques of decoupling high-speed digital systems can also be applied to RF systems. RF systems are often easier to decouple. Most RF circuits create supply variations in very narrow frequency ranges. In such applications, a decoupling capacitor can be chosen such that the resonance of the capacitor (including the PCB connections) occurs at the frequency of the circuit. Furthermore, most RF circuits only require a high-frequency decoupling impedance of several ohms. In other words, they are a lot more tolerant of supply variations.

POWER PLANE RIPPLES

You may wonder what happens when high-frequency decoupling caps are not placed locally? The effect is that of a circular wave that propagates away from the IC variation. Because the PCB is electrically long at these higher frequencies, the ripples propagate like ripples on a pond. If a decoupling cap is placed near the source, most of the energy can be absorbed. If the decoupling cap is placed far from the source, it can successfully shunt only some of the wave to ground.

90 DEGREE TURNS AND CHAMFERED CORNERS

A cause of misconception and unnecessary worry for many designers is the question of whether or not 90 degree turns affect transmission line

Figure 12.13 90 degree turns on PCB transmission lines. The uncompensated 90 degree turn (left) is satisfactory below 1 GHz. For higher frequencies, it may be necessary to compensate for the bend by chamfering the corner (right).

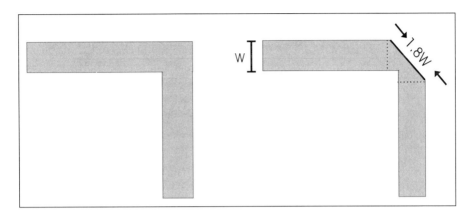

behavior and radiation. Unless you are working at frequencies (or equivalent edge rates) above 1 GHz, the answer is an emphatic no. Furthermore, unless you are working at frequencies above 10 GHz, the effect is still small. The major effect of a bend is to add a parasitic capacitance to the transmission line at that point. The parasitic value can be between 0.001 pF to 0.2 pF, depending on the geometry and characteristic impedance of the line. Compared to the discontinuities presented at vias, IC pads, and connectors, the parasitics of 90 degree turns are certainly minor. If you are designing at GHz frequencies, you can use the technique of chamfered corners (illustrated in Figure 12.13) to minimize the parasitic capacitance.

LAYOUT OF TRANSMISSION LINE TERMINATIONS

When using terminations on transmission lines, the placement of the termination is very important. For a series termination, the terminating resistor is placed in series with the driving source. The component must be placed within a critical distance of the IC output pin. The critical distance is the distance that it takes that signal $1/3$ t_{rise} to travel on the transmission line, where t_{rise} is the rise time of the driving IC. For good designs, you should strive to meet $1/6$ t_{rise} ("How close is close enough," H. Johnson, 1998). For a parallel termination, the component is placed in parallel with the load. The termination component should be placed

Figure 12.14 Distribution of digital clock signals. A bad layout and good layout are shown. The best results are obtained when wiring is point-to-point.

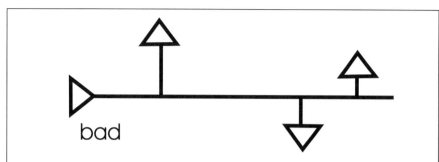

A clock signal is routed using a single bus. Each receiving circuit is connected using a fairly long stub ("off ramp"). The topology is not recommended because each stub appears as a capacitance and will cause a reflection on the line. Note also that there is an unconnected stub at the end of the line. This stub will also cause reflections.

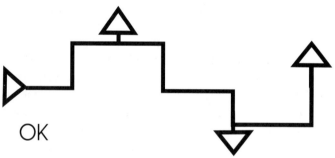

The clock is routed in a true "daisy chain" fashion. The clock signal is brought directly to each receiving circuit. This topolgy minimizes the reflections at each receiver.

after the last load on the transmission line. The termination should never be placed before any load. Furthermore, when several loads are placed on the same transmission line, the technique of daisy chaining should be used instead of the technique of "off-amps." Figure 12.14 illustrates this concept. Of course, routing clock lines individually is the best solution. Terminating both ends of a transmission line is the best technique, because reflections from any discontinuities along the line are absorbed at both source and load.

ROUTING OF SIGNALS: GROUND PLANES, IMAGE PLANES, AND PCB STACKUP

The ideal design for high-speed signals is as follows:

- The outer layers have a ground directly underneath forming microstrip traces for all outer layer signals
- Every internal signal layer is surrounded by a ground above and below, forming the ideal strip-line geometry
- Signals traces have a maximum of two vias; one via at the source IC pin and one at the destination IC pin. In between these two pins, the signal never changes layers

It is a good idea to strive for these ideals, especially on the most critical signals, but in practice, many compromises must be made. *When routing the layout of any high-speed PCB, it is a good practice to route the most critical signals (clocks, interrupts, high-current lines, very fast signals, etc.) first, using the most direct routes with good ground reference and minimal vias.* Once the critical signals are handled, compromises can then be made with the less problematic signals.

Often alternating signal layers are reserved for horizontal routes and vertical routes. This not only helps the job of the autorouting software, but it also reduces coupling of signals on adjacent layers because the traces are mostly orthogonal.

Because every PCB layer adds cost to the product, image planes are often used for digital boards. An *image plane* is a power plane that serves as a virtual ground for the adjacent transmission lines. High-speed signals are by definition AC signals, and if good decoupling is used, the ground and power planes will be effectively shorted together at AC frequencies. The power plane can therefore serve as the return path for AC signals to follow. Because the impedance is never zero between the planes, and because the current has to eventually find its way to the ground pins of the ICs, image planes can never perform as well as true ground planes.

One golden rule that should never be broken is: *Never place more than two signal layers between planes.* In other words, all signal traces should have a ground or power plane directly adjacent above and/or below. For example, if you placed three signal layers between two ground planes, the signals on the middle layer would couple equally to each ground plane with their fields extending through both of the other signal layers. Uncontrollable cross-coupling will occur between the signal layers. Two signal layers between planes is referred to as *dual* or *asymmetric strip-line*. One signal layer between planes is of course just called strip-line.

The definition of how many layers are used in a PCB and how each layer is utilized (e.g., ground, power, signals) is referred to as the *PCB stackup*. You can find suggestions for PCB stackups in Montrose (2000) and Johnson (1993).

3W RULE FOR PREVENTING CROSSTALK

Any two transmission lines on the same PCB layer are going to exhibit a certain amount of coupling to one another, especially if they are parallel. To help prevent errors due to coupling, the "3W" rule can be followed: The distance between the centers of two traces should be at least three times the width of the traces. Stated more simply, the space between two traces should be at least twice the width of each trace.

LAYOUT MISCELLANY

When making measurements of digital signals, be sure that your oscilloscope has enough bandwidth. The oscilloscope's bandwidth should be at least $0.5/t_{rise}$.

Above 1 GHz, the performance of FR4 PCBs becomes a problem due to dispersion. Dispersion is caused by loss in the FR4. All dielectric materials have some loss, which appears as conductance between the metal layers. This conductance affects the characteristic impedance and velocity of the transmission lines at high frequencies. The result is that signals of different frequencies travel at different speeds. The edges of digital signals contain many frequencies, and when subject to dispersion, the edge is lengthened and possibly distorted. The same effect can occur in RF systems. RF signals consist of a carrier wave modulated by the information signal (voice, video, data, etc.). This modulation typically produces a signal with a small, but non-zero bandwidth about the carrier frequency. Dispersion thus causes distortion of RF signals as well. Several alternative (and more expensive) materials are available for GHz PCBs. These materials exhibit less loss than FR4, allowing for higher-frequency operation. Rogers Corporation (www.rogers-corp.com) specializes in these materials and provides a wealth of information on their web site.

Many PCBs have ground planes on several layers. Sometimes there is one internal ground layer, and then most of the outside layers are grounded. Other PCBs may contain more than one internal ground layer. In either case, the separate grounds must be properly connected for high-speed operation. Unless the grounds are isolated for a reason (different circuit functionality), the grounds must be connected in a distributed manner across the PCB. The "$\lambda/20$" rule should be followed in these applications: The vias that stitch the ground planes together should be placed at a high enough density such that no vias are more than $\lambda/20$ distance apart, where λ is the wavelength of the highest frequency (or equivalent edge rate) in the design.

To learn more about the tricks of the trade of digital PCB layout (e.g., PCB layer stackup suggestions, edge rates for logic families), I recommend the two companion books by Montrose. Montrose (2000) focuses on PCB layout techniques and Montrose (1999) covers EMC and EMC as it applies to PCB design. Johnson (1993) is another very good book that covers the analog subtleties of high-speed digital design.

LAYOUT EXAMPLES

Probably the best way to learn good layout techniques is to learn by example. The following figures (Figure 12.15 through 12.21) show examples of bad and good layouts for several circuits.

PART III: CABLING

Before learning about cabling, it is important to discuss the topics of ground loops, differential mode radiation, and common mode radiation.

GROUND LOOPS (MULTIPLE RETURN PATHS)

Ground loops occur when there are multiple pathways that ground currents can take.

Ground loops are usually only a problem when the loops are large, like those that can occur when connecting two separately powered products. Figure 12.22 shows two products connected by a two-conductor signal path.

Both of the products are powered by mains (wall socket) voltage, and hence are referenced to Earth. One of the conductors of the interface is connected to ground at each of the products. If a signal is sent across the cable, it can return by the second wire, as intended, or it can return

Figure 12.15 Simple changes can make a big difference with high-frequency signals. Here are two examples of routing the ground to an IC. In the first example, the ground currents are forced to traverse a rather large loop to get from the IC to the ground plane underneath the traces. The second example provides much shorter ground paths.

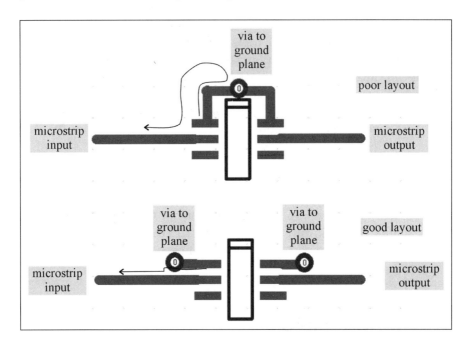

via the ground connection (i.e., through the power supply ground wire and through wall wiring to the other power cable and into other products). This second return path causes a ground loop to be formed, as shown in the figure. At low frequencies (under 1 MHz is a decent rule of thumb), some of the return signal will follow the intended path and some will follow the longer path. The current that follows the longer path creates a rather large loop. This large loop creates large stray fields, and can create large radiated fields. At higher frequencies, the large path presents too high an impedance (due to its inductance) for any current to follow it; high-frequency current will follow along the intended return path, close to the signal current. This is one of the few exceptions to the general rule that high-frequency signals are more problematic than low-frequency signals.

Another problem associated with ground loops is that of susceptibility. The large loop area in conjunction with small loop resistance effi-

Figure 12.16 Here are two examples of routing a signal trace and its return to a connector. In the first example, the current forms a rather large loop. In the second example, the current loop is small.

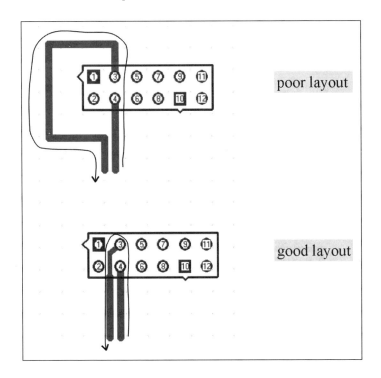

ciently captures the ambient magnetic fields. This loop can also serve as a loop antenna for receiving radiated fields. For coaxial cables, this problem is again alleviated at high frequencies for two reasons. First, when the frequency of the interfering signal is high, the outer conductor of the coaxial cable becomes comparable in thickness to the skin depth of the metal. Thus, the signal return current flows on the inside of the outer conductor, and the interference current flows on the outside of the outer conductor. Interference between the two circuits is greatly reduced because it is as if the return current and interference currents are flowing in separate circuits, like a triaxial cable (refer to the section on cables). Second, the mutual inductance of the inner and outer conductors presents a large impedance to high-frequency common mode currents, making it difficult for high-frequency ground loop currents to flow. In summary, ground loops are mostly a low-frequency problem,

Figure 12.17 Two examples of power-supply decoupling on a two-layer board. Light gray denotes bottom-layer traces. The first layout creates a large loop for the decoupling path, which will provide poor performance for high-speed signals. The second layer is better because the decoupling loop is reduced.

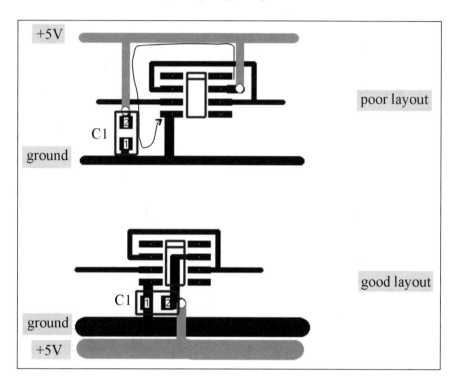

causing noise in audio systems and other low-frequency applications, including control systems and sensor measurements.

DIFFERENTIAL MODE AND COMMON MODE RADIATION

Unintentional radiation can be described as either differential mode or common mode. The two terms relate to the currents that create the radiation. Differential mode currents are equal but travel in opposite directions. Common mode currents travel in the same direction.

On any transmission line, the signal current and the return current travel in opposite directions. As long as the two currents are close together the radiation is very small, albeit non-zero. The two currents

Figure 12.18 Here are two layout examples of a switch-mode power supply. All vias connect to the ground plane. The first layout contains large loops. Therefore, the large currents that cycle through the capacitor, C1, may cause large amounts of radiation. The second layout shrinks the loops, uses shorter, fatter traces, and uses multiple vias to reduce parasitic inductance.

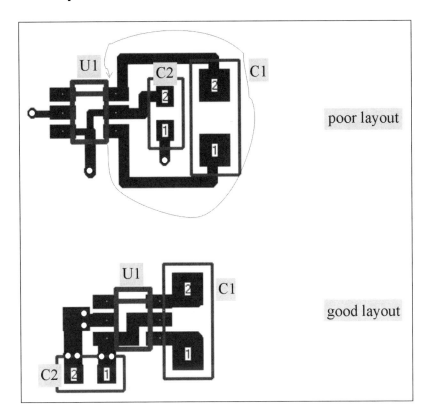

form a differential pair of currents with a space between them. The radiation of the line is proportional to the cross-sectional area of the space in between. Such radiation is like that of a loop antenna and is magnetic in the near field. For a microstrip or strip-line transmission line, the cross-sectional area is very small, equal to the thickness of the dielectric times the length of the line. However, at ICs and connectors the signal and return may form much larger loops, especially if care is not given to how the connections are laid out.

Common mode radiation arises when there is a net current traveling in one direction on a transmission line. The extra current may be trav-

Figure 12.19 Examples of connecting a decoupling capacitor to the power and ground planes. Short, wide traces with multiple vias produce the lowest parasitic inductance. The example on the far right is probably overkill.

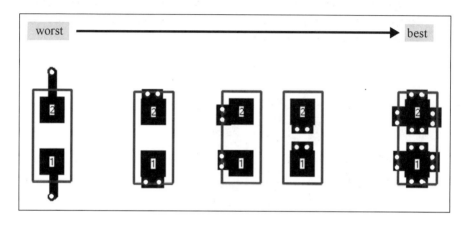

Figure 12.20 Examples of IC decoupling. In the first example, the power and ground traces only connect to the planes with a single via, and the decoupling path is rather large. In the second example, multiple vias are used on every component, and the decoupling capacitors are placed underneath the IC on the bottom layer.

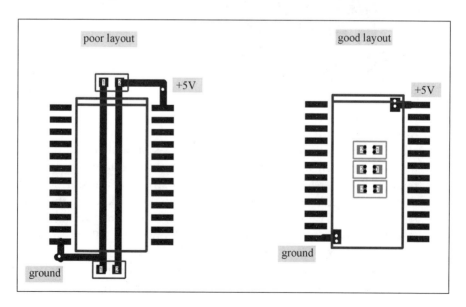

Figure 12.21 Examples of ground-layer routing under RF microstrip transmission lines. The light gray represents the ground plane. The dark gray represents traces on the ground plane. In the first example, the DC signal trace travels a long distance on the ground plane, blocking the flow of high-frequency current. In the second example, two small ground plane traces are used to limit problems.

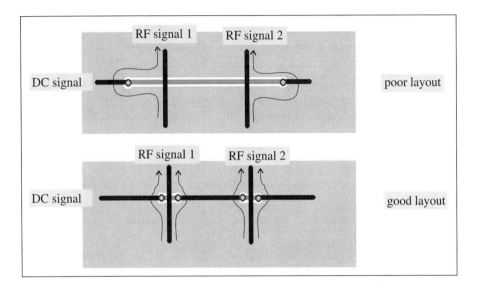

Figure 12.22 Demonstration of a ground loop. A signal travels from one instrument to another over a coaxial cable. Some of the return current travels on the cable shield, and some of the return current travels through a ground connection via the power cabling and mains wiring. The two return paths form a large loop that can easily pick up noise, which couples into the equipment.

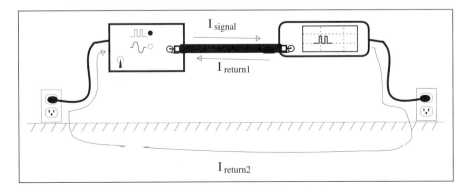

eling on both conductors or on just one conductor. The common characteristic is that the addition of the current on the signal and return conductors does not sum to zero. Such currents are referred to as asymmetric currents. This result is never intended, but arises when return currents take alternative pathways back to the source. For example, consider a cable between two products that are mains powered. If 1 amp of signal current travels down the cable, and 0.1 amps return via the ground, then only 0.9 amps return via the second conductor in the cable. In other words, a ground loop has occurred. The net current on the cable is $1.0 - 0.9 = 0.1$ amps. This common mode current translates to radiation. Although in this case, it is really loop radiation because the current follows a loop through the alternative ground. Ground loops only tend to occur at low frequencies.

Another way current can bypass the second conductor of the cable is through parasitic capacitive coupling. This type of return current can occur with cables and with PCB traces. Figure 12.23 illustrates the concept.

Some of the current returns through the parasitic capacitor, leaving a net current on the cable. In this situation, the cable acts like a whip (monopole) antenna. As such, it can create tremendous amounts of radiation if associated with high frequencies. The common mode problem is exacerbated when a transmission line includes local regions with large loops. The loop presents an inductive impedance which increases with frequency. As frequency increases, the transmission line loop becomes larger in impedance and the parasitic capacitive pathways become smaller in impedance. The parasitic pathways then becomes easier paths for return current. Thus, a local loop in a transmission line causes two problems: 1) the loop itself creates differential mode radiation; 2) the inductance of the loop can force current to return via parasitic capacitance, resulting in common mode radiation on the rest of the transmission line.

One method of thwarting common mode currents is to use a so-called *common mode choke*, as shown in Figure 12.24. A common mode choke consists of a transmission line wound into a coil or wrapped around a ferromagnetic core. Although the schematic symbol looks like a generic transformer, for proper performance the two wires should be wound around the core in tandem as a transmission line. Such wiring is not typical for generic transformers. The operation of a common mode choke is similar to the way a transformer works. The device presents a series inductance in both conductors, and the two inductors are well coupled. If a current travels on the one conductor and an equal but opposite current travels on the second conductor, the

Figure 12.23 Common mode currents arise due to asymmetries in circuits. The circuit asymmetries give rise to asymmetric parasitic paths, leaving a net current on the cable. The net common mode current acts as an efficient antenna.

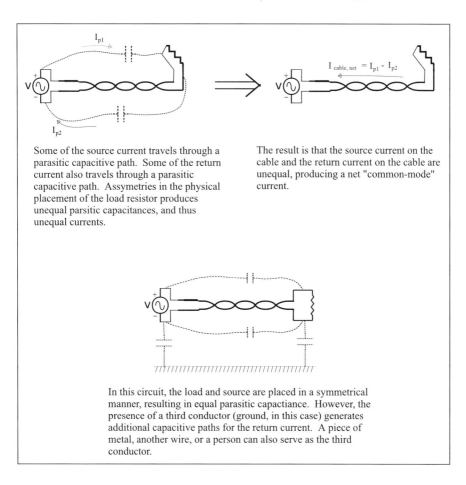

Some of the source current travels through a parasitic capacitive path. Some of the return current also travels through a parasitic capacitive path. Assymetries in the physical placement of the load resistor produces unequal parsitic capacitances, and thus unequal currents.

The result is that the source current on the cable and the return current on the cable are unequal, producing a net "common-mode" current.

In this circuit, the load and source are placed in a symmetrical manner, resulting in equal parasitic capacitance. However, the presence of a third conductor (ground, in this case) generates additional capacitive paths for the return current. A piece of metal, another wire, or a person can also serve as the third conductor.

magnetic fields of each current cancel each other, resulting in zero magnetic field. The inductance disappears for differential currents. An asymmetric current, on the other hand, does not produce this cancellation effect, and is presented with the high inductance of the coil. The result is that differential currents pass through unchanged, but common mode currents are "choked" by the inductance. Common mode chokes are commonly used on the power cords inside most products, to prevent common mode currents on the cord. They are also used on most computer CRT cables. Take a look at the end of any computer monitor cable.

Figure 12.24 A common mode choke is constructed in the same manner as an inductor, except instead of winding a single wire, the common mode choke is wound using a transmission line.

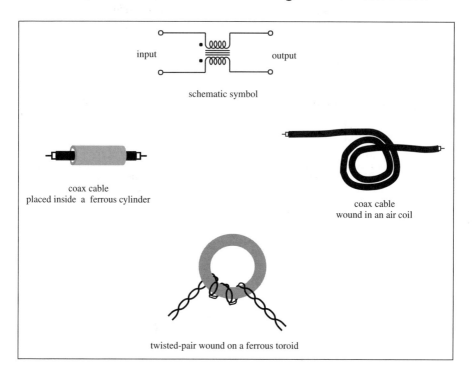

About an inch or so from the end connector, you will notice that the cable has a cylindrical lump in it. Inside this lump is a common mode choke.

CABLE SHIELDING

Because cables are typically the longest component of many electronic systems, they are also the most prone to radiate and most susceptible to electromagnetic interference. Proper cable shielding can make a difference of close to 80 dB in some cases, so it is a topic well worth spending the time to understand. There are two basic cabling types that provide shielding: coaxial cables and twisted wires pairs. *Coaxial shields provide an inherent immunity to electric field interference, and twisted pairs provide an inherent immunity to magnetic field interference.*

CABLE SHIELDING

The worst type of cabling is that of a single wire, using the ground for return. The Earth's impedance can vary greatly, and the large loop area involved makes these systems horribly noisy. The Earth return system was used for telegraph and telephone systems of the late 1800s and early 1900s, but it was soon realized that they were much too susceptible to crosstalk and interference from manmade and natural noise sources. Crosstalk may not have been a problem when there was only one telegraph or telephone cable in each town and no radio stations, but as the electronic age was ushered in, engineers switched to using two-conductor transmission lines.

Coaxial Cables Below the Break Frequency

To provide excellent electric field shielding, a metal cylinder can be placed around the signal conductor and used for the return current. This geometry forms the common coaxial cable. The term *coaxial* refers to the fact that the wire (inner conductor) and the shield (outer conductor) share the same central axis. In other words, the wire is placed in the exact center of the shield. For the coaxial cable to work effectively, it must surround the inner conductor completely, without any gaps or slots. The shield must be chassis-grounded to operate as intended (otherwise, electric fields will capacitively couple to the outside of the shield and then capacitively couple to the inside of the shield). When the shield is grounded, the electric fields couple to the shield and then are shunted to ground. Some interference still occurs because of the finite resistance of the shield, but the interference is greatly diminished. For electric shielding, the shield needs to be connected to the chassis-ground of the equipment at only one side. However, if the cable is electrically long, the grounding needs to be performed in multiple places; the ground connections should be placed every $1/20\lambda$ to $1/10\lambda$ or less.

The shield of a coaxial cable unfortunately provides no inherent magnetic shielding. However, there are techniques of grounding that can provide excellent magnetic shielding with coax. The most important factor in successful cabling is to provide a path for return current to flow through the shield, where it is close to the signal conductor. By following this rule, most of the current returns in the shield and not the earth. The technique allows the signal loop to be small, preventing radiation. Ideally, the shield should be connected to chassis-ground at only one location so that ground loops cannot form. The rules for the treatment of coaxial cables at *low frequency* (below the cable break frequency) are as follows:

- The shield (outer conductor) should be connected to the signal common at both ends.
- If the signal at a cable end is referenced to ground (earth) through a resistor or circuit, the shield should also be grounded at that point.
- If the signal at a cable end is not referenced to ground (earth) through a resistor or circuit, the shield should not be grounded at that point.
- Ideally, the signal common and shield should be grounded (earthed) at exactly one location. Grounding at two locations is acceptable, but reduces magnetic performance.
- Electrically long cables are an exception to the rule, and should be grounded at least every 1/10 λ.
- If the shield is grounded at both ends, a common mode choke can be used to prevent common mode currents. The use of transformers or optocouplers will also break ground loops and prevent common mode currents.
- A simple way to create a common mode choke is to coil a portion of the cable into a loop. This technique is commonly used periodically along the utility poles in CATV distribution.

Triaxial Cables and Ferromagnetic Shielding

To provide even better shielding effectiveness, a second shield can be used. This second shield allows ground current to circulate on the outer shield and signal return current to flow on the inner shield. This arrangement prevents common impedance coupling between the interference current and the signal current in the shield and provides the best performance. Such shielding is used for very sensitive measurements. The outer shield should be grounded via the equipment chassis at both ends. The inner shield is connected to the signal common at both ends, which may or may not be referenced to the chassis ground. Another technique is the use of an added shield made from a ferromagnetic material such as mu-metal. The ferromagnetic material concentrates any ambient magnetic fields, reducing the penetration of the field into the interior of the cable.

Break Frequency

The division between high and low frequencies on cables is determined by the *break frequency* of the cable. The break frequency (also called the cutoff frequency) is simply a low-pass filter effect caused by the shield inductance and shield resistance, $f = R/2\pi L$, where R and L are the series resistance and inductance of the shield, respectively. Cables behave *much* differently below the break frequency, as mentioned earlier in the chapter. At high frequency, the inductance prevents ground loop currents and other types of common mode currents. Above the break frequency, the characteristic impedance of the cable is approximately

$$Z_o = \sqrt{\frac{L}{C}}$$

Below the break frequency, the characteristic impedance is approximately

$$Z_o \approx \sqrt{\frac{R}{i2\pi f C}}$$

Refer to Chapter 8 for the exact formula. Note that below the break frequency, Z_o is a complex number with a phase angle of 45 degrees. The break frequency for typical coaxial cables falls in the range between 500 Hz and 10 kHz.

Coaxial Cables Above the Break Frequency

Surprisingly, above the break frequency, coaxial cable performance actually improves. At about 5 times the break frequency, virtually all the current travels in the shield and none travels via ground loops. Thus, the guideline for grounding high-frequency coaxial cables is simple.

At high frequencies (five times the break frequency) coaxial cable should be grounded at both ends.

Skin Depth and Transfer Impedance

As frequency increases, the associated skin depth becomes very small; when the skin depth is 1/5 or less than the thickness of the outer conductor, a coaxial cable behaves like a triaxial (double shielded) cable. Signal current flows exclusively on the inner surface of the shield, and induced interference current flows on the outer surface of the shield. For most shields, this change occurs at about 1 MHz. The measure of isolation created by the skin effect is often expressed in terms of *transfer impedance*. At DC, the transfer impedance is just the DC resistance of the cable. Once the skin depth becomes less than the thickness of the shield, the transfer impedance quickly drops to zero. The voltage that develops in a cable because of induced interference currents in the shield is proportional to the transfer impedance. Thus, as the transfer impedance goes to zero, so does the interference voltage.

At high frequencies (>1 MHz, typical) a coaxial cable with solid shield performs like a triaxial cable.

Solid and Braided Shields

Coaxial shields come in two types—solid and braided. A solid shield is just what it sounds like—a solid conductor that completely surrounds the inner conductor. Aluminum is most often used for solid shields. Aluminum cannot be soldered to, so crimp style connectors must be used with this type of cable. For CATV distribution, the aluminum is an extruded/bonded piece of metal. This shielding is ideal. For most other applications, the shield is a piece of aluminum foil. The foil typically has a crease in it, which prevents current from circulating and can cause high-frequency problems.

Braided coax utilizes a copper braid for the shield. Because no braid can provide complete coverage, the shielding properties are compromised. The most dramatic effect is the behavior of the transfer impedance. *At high frequencies, the transfer impedance actually increases instead of decreasing because of the inductance from the holes in the braid. Thus the magnetic shielding behavior becomes worse instead of better.* Braided shields can be the source of strange problems such as GHz oscillations. The best high frequency solution is a cable with both an aluminum foil shield and a copper braid. Such cables are available from Belden (www.belden.com) and can exhibit transfer impedances as low as 1 mΩ/meter at 500 MHz.

Ribbon Cables

Ribbon cables are very ineffective at providing any type of shielding, and should not be used except inside products over very short distances. (Lab use of ribbon cable for RS-232 signaling is also often acceptable, but don't intend to sell any products based on external ribbon cabling.) When using ribbon cables, it is a good idea to use a separate ground wire for every signal wire. The wiring should be such that signal wires and ground wires alternate in succession. This technique reduces inductive and common impedance coupling between the wires.

CABLE SHIELDING

Twisted Pair Cables

Twisted pair cable consists of a pair of wires twisted consistently to achieve a constant characteristic impedance. The twisted pair cable is the complement to the coaxial cable in that it provides inherent magnetic shielding but no electric shielding. The magnetic shielding arises from the twisting. Each twist reverses the polarity of the loop; therefore magnetic fields can only couple to the small loop twists and not the cable as a whole. The shielding effectiveness increases with finer pitch (more twists per inch). Finer pitch also allows for higher-frequency operation because it prevents radiation better. Higher-frequency operation also requires that the variation in the twisting pitch (i.e., the pitch tolerance) must be minimized. Twisted pair cable is used for carrying phone signals and for twisted pair Ethernet local area network (LAN) signals. Standards exist for different types of cable. Category 1 cable is specified for voice signals; category 3 cable has a finer pitch and is specified for 10 Mbit/sec signaling; category 5 is specified for 100 Mbit/sec signaling; category 6 cable is specified for 250 Mbit/sec cabling; category 7 cable is planned for 600 Mbit/sec operation. Several pairs can be used in parallel to achieve higher bit rates. These cables are specified to transmit data up to 100 meters. The high-frequency characteristic impedance is 100 ohms.

There is also a break frequency for twisted pair cables. The break frequency for typical twisted pair wires, like those used in telephone wiring, falls in the range between 10 kHz and 100 kHz. At audio frequencies the magnitude of characteristic impedance is typically specified as 900 ohms or 600 ohms, in accordance with telephone standards. The line is matched to a 600 ohm source at the telephone company's central office (CO). At 60 Hz it rises to 10 kohms or more. This is in drastic contrast to the purely real, 100 ohm, high-frequency characteristic impedance typical of twisted pairs.

There are two methods to provide electric field shielding on twisted pairs. The first method is to use a metal shield around both of the pairs. This shield is grounded via the equipment chassis, and it must be grounded at either one end or both ends. However, several options exist for grounding the shield. For low frequencies (below break frequency), grounding at one end is best. For high frequencies, grounding at both ends is best. One variation is to ground the cable at one end and then connect the other end to ground via a capacitor. This technique provides a single ground for low frequencies, and two grounds for high frequencies. The capacitor should have a high voltage rating (to withstand

ESD events), and a large value resistor should be used in parallel to further drain any static charges.

Single-Ended versus Balanced Signaling*

So far all the cable techniques I have discussed assume that one of the conductors is connected to the signal common. In the case of coaxial cables, the outer conductor is always connected to the signal common (and possibly ground). Because twisted pair is a symmetric transmission line, it doesn't matter which conductor is connected to ground. This method of cabling is referred to as single-ended cabling.

A better way to send signals is to use balanced signaling. In a balanced system, each wire has the same impedance to the circuit common (which is typically grounded). This effect can be achieved by using a center-tapped transformer with the center tap connected to ground, or by using differential amplifiers that drive a positive signal on one wire and an equal but opposite voltage on the other signal. For proper balancing, the output impedance of each amplifier output must be equal. At the receiving end, the difference in voltage between the signals is measured. The receiving end circuit may or may not be referenced to ground.

This technique has many great benefits. First, it is immune to ground differences at the sending and receiving ends, because only the difference of the voltage between the wires is measured. The common voltage to ground is irrelevant. Second, the balancing provides immunity to electric fields. If placed in an electric field, currents may be induced, but due to balancing, the currents and voltages are equal in the two wires. Consequently, an equal interference voltage is added to both wires. Since the receiver amplifies the difference, the received signal is unaffected.

Telephone lines use balanced twisted pair signaling. At the central office, each wire is typically connected to ground through 300 (or 450) ohms of resistance. This arrangement provides balancing and impedance matching to 600 (or 900) ohms. The receiving end (your telephone set) is typically floating with respect to ground. Because telephone cables can be several miles long and often travel underneath power lines, shielding is extremely important. The measurement that quantifies how well a twisted pair is balanced is called *longitudinal balance*. Any mismatch in impedance to ground reduces the electric field immunity. Twisted pair Ethernet, as well as data protocols such as RS-422, also utilizes balanced twisted pair signaling. The original Ethernet standard

---

*Balanced signaling requires a symmetric transmission line, such as twisted pair. Coaxial cable must be single-ended.

specified coaxial cable, but twisted pair is now exclusively used in the Ethernet standards. The main reason for the change of technologies is the ease of physically routing twisted pair as compared to coaxial cable, which is bulkier and harder to bend.

Cabling Summary

Twisted pair cable works better than coaxial cable for low impedance (impedance less than the characteristic impedance of the cable) since it is better at preventing magnetic coupling. (Recall that low-impedance lines are mostly susceptible to magnetic field coupling.) For high-impedance applications coaxial cable provides better performance than unbalanced twisted pair. However, a well-balanced twisted pair can perform equally well as a coaxial cable. Shielding a balanced twisted pair provides extra immunity. For low-frequency applications, such as audio, shielded balanced twisted pair is the best solution. Thanks to the category standardization, data is very well served by balanced twisted pair cabling. Analog RF and microwave is the domain of coaxial cables. Some coaxial cabling systems (cables and connectors) can support frequencies of 40 GHz and higher.

BIBLIOGRAPHY AND SUGGESTIONS FOR FURTHER READING

Agilent Technologies, *Making Precompliance Conducted and Radiated Emissions Measurements with EMC Analyzers,* Agilent Technologies, AN 1328.

Breed, G. A., "Notes on Power Supply Decoupling," *RF Design,* October 1993.

Breed, G. A., "Notes on High-Speed Digital Signals," *RF Design,* December 1994.

Breed, G. A., "Initial Guidelines for Layout of Printed Circuit Boards," *RF Design,* September 1995.

Breed, G. A., "EMI/RFI Radiation and Susceptibility from Cables and Enclosures." *Applied Microwave Wireless,* January 2000.

Brooks, D., "90 Degree Corners: The Final Turn," *Printed Circuit Design Magazine,* January 1998.

Brooks, D., "Bypass Capacitors: An Interview with Todd Hubing," *Printed Circuit Design Magazine*, March 1998.

Brooks, D., "Differential Impedance. What's the Difference," *Printed Circuit Design Magazine,* August 1998.

Cain, J., "The Effects of ESR and ESL in Digital Decoupling Applications," AVX Corp., March 1997.

Cain, J., "Interconnect Schemes for Low Inductance Ceramic Capacitors," AVX Corp.

Caldwell, B., and D. Getty, "Coping with SCSI at Gigahertz Speeds," *EDN*, July 6, 2000.

Carsten, B., "Sniffer Probe Locates Sources of EMI," *EDN*, June 4, 1998.

Degauque, P., and J. Hamelin, *Electromagnetic Compatibility*, Oxford: Oxford University Press, 1993.

Fairchild Semiconductor, *Designing with TTL*, Fairchild Semiconductor, AN-363, June 1984.

Fairchild Semiconductor, *Follow PC-Board Design Guidelines for Lowest CMOS EMI Radiation*, Fairchild Semiconductor, AN-389, January 1985.

Fairchild Semiconductor, *Fact Design Considerations*, Fairchild Semiconductor, AN-MS-539, November 1988.

Fairchild Semiconductor, *Fast Design Considerations*, Fairchild Semiconductor, AN-661, March 1990.

Fairchild Semiconductor, *Design Considerations*, Fairchild Semiconductor, AN-MS-520, August 1993.

Fairchild Semiconductor, *Terminations for Advanced CMOS Logic*, Fairchild Semiconductor, AN-610, February 1998.

Fairchild Semiconductor, *Understanding and Minimizing Ground Bounce*, Fairchild Semiconductor, AN-640, 1998.

Fairchild Semiconductor, *Considerations in Designing the Printed Circuit Boards of Embedded Switching Power Supplies*, Fairchild Semiconductor, AN-1031, April 30, 1999.

Fiore, R., "Capacitors in Broadband Applications," *Applied Microwaves and Wireless*, May 2001.

Foo, S., "High-frequency Laminates," *RF Tutorial*, April 1998.

Gerke, D., "RFI Still Threatens Computers," *RF Design*, June 1998.

Gerke, D., and W. Kimmel, "EMI and Circuit Components: Where the Rubber Meets the Road," *EDN*, September 1, 2000.

Greb, V. W., and C. Grasso, "Don't Let Rules of Thumb Set Decoupling-capacitor Values," *EDN*, September 1, 1995.

Greim, M. C., "High-End Digital Systems Give a Thumbs Down to Rules of Thumb," *EDN*, June 5, 2000.

Haller, R. J., "The Nuts and Bolts of Signal-Integrity Analysis," *EDN*, March 16, 2000.

Hardin, K. B., *Decomposition of Radiating Structures to Directly Predict Asymmetric-Mode Radiation*, University of Lexington Ph.D. Dissertation, 1991.

Hartley, R., "Taking a Bite out of EMI," *Printed Circuit Design*, January 2000.

Hewlett-Packard, *Designing for Electromagnetic Compatibility—Student Workbook*, California: Hewlett-Packard, Course No.-HP 11949A, 1989.

INTEL, *Pentium III Xeon Processor Power Distribution Guide*, Document Number: 245095-001, March 1999.

INTEL, *Intel Pentium 4 Processor/ Intel 850 Chipset Platform Design Guide*, Document Number 298245-001, November 2000.

INTEL, *Intel Researchers Build World's Fastest Silicon Transistors*, Press Release, INTEL Corporation, 2001.

INTEL, *Controlled Impedance Design and Test*, INTEL Corp.

Johnson, H., *High-Speed Digital Design: A Two-Day Workshop in Black Magic*, Redmond, Wash.: Signal Consulting, 1997.
Johnson, H., "Mutual Understanding," *EDN*, January 1, 1998.
Johnson, H., "How Close Is Enough," *EDN*, April 9, 1998.
Johnson, H., "Power-plane Resonance," *EDN*, September 1, 1998.
Johnson, H., "Who's Afraid of the Big, Bad Bend?" *EDN*, May 11, 2000.
Johnson, H., and M. Graham, *High-Speed Digital Design: A Handbook of Black Magic*, Englewood Cliffs, NJ: Prentice-Hall, 1993.
Keithley Instruments, *Model 6514 System Electrometer Instruction Manual*, 6514-910-01 Rev. C, Cleveland: Keithley Instruments, Inc., 1998.
Kimmel, W., and D. Gerke, *Designing for EMC—Practical Tools, Tips, and Techniques for Bullet Proof Designs*, St. Paul, Minnesota: Kimmel Gerke Associates, 1997.
Liou, D., *Controlled Impedance in PCB Design*, Nortel Networks Internal Document, October 20, 1998.
Lipman, J., "Tools and Techniques Stifle EM Emissions," *EDN*, November 6, 1997.
Lipman, J., "Board-Level Signal-Integrity Analysis: Sooner Is Better," *EDN*, July 16, 1998.
Marsh, D., "Category 6 Cable: Gigabit Ethernet Over Copper," *EDN*, December 9, 1999.
Martin, A. G., and R. K. Keenan, *Improved Noise Suppression via Multilayer Ceramic Capacitors (MLCs) in Power-Entry Decoupling*, AVX Corp., S-MINS3942.5M-R.
Montrose, M. I., *EMC and the Printed Circuit Board—Design, Theory, and Layout Made Simple*, New York: IEEE Press, 1999.
Montrose, M. I., *Printed Circuit Board Design Techniques for EMC Compliance—A Handbook for Designers*, 2nd Edition, New York: IEEE Press, 2000.
Morgan C., and D. Helster, "New Printed-Wiring-Board Materials Guard Against Garbled Gigabits," *EDN*, November 11, 1999.
Morrison, R., *Grounding and Shielding Techniques*, 4th Edition, New York: John Wiley, 1998.
Novak, I., *Powering Digital Boards Distribution and Performance*, Meeting of Greater Boston Chapter IPC Designer's Council, February 9, 1999.
Novak, I., "Reducing Simultaneous Switching Noise and EMI on Ground/Power Planes by Dissipative Edge Termination," *IEEE Tr. CPMT*, Vol.22, No.3, August 1999.
Novak, I., *Accuracy Considerations of Power-Ground Plane Models*, 1st Topical Meeting on Electrical Performance of Electronic Packaging, San Diego, Calif.: SUN Microsystems Inc., October 1999.
Novak, I., *Measuring Milliohms and PicoHenrys in Power Distribution Networks*, SUN Microsystems Inc., February 2000.
Ott, H. W., *Noise Reduction Techniques in Electronic Systems*, 2nd Edition, New York: John Wiley, 1988.
Ott, H. W., "Partitioning and Layout of a Mixed-Signal PCB," *Printed Circuit Design*, June 2001.
Paul, C. R., *Introduction to Electromagnetic Compatibility*, New York: John Wiley & Sons, 1992.

Pease, R. A., *Troubleshooting Analog Circuits*, Newton, Mass.: Butterworth–Heinemann, 1991.

Pham, N., *System Board Power Distribution Guidelines for Mach5*, IBM Corp., AN-022, September 1997.

Prymak, J. D., *PE Series Capacitors Decoupling and/or Filtering*, AVX Corp., S-CD00M301-R.

Roy, T., L. Smith, J. Prymak, *ESR and ESL of Ceramic Capacitor Applied to Decoupling Applications*, SUN Microsystems, Inc., MS MPK15-103.

Seltzer, G. S., R. Ahy, M. Razmi, " Electromagnetic Theory Imposes Constraints on PC-Board Design," *EDN*, July 21, 1994.

Smith, L. D., *Decoupling Capacitor Calculations for CMOS Circuits*, IBM Corporation, F32/61C.

Smith, L. D., *Simultaneous Switch Noise and Power Plane Bounce for CMOS Technology*, SUN Microsystems Inc., MS MPK15-103.

Smith, L., R. Anderson, D. Forehand, T. Pelc, T. Roy, *Power Distribution System Design Methodology and Capacitor Selection for Modern CMOS Technology*, MS MPK15-103, SUN Microsystems Inc.

Sugasawara, D., and J. Paetau, *Safety, EMI and RFI Considerations*, Application Note: 42007, Fairchild Semiconductor, June 1996.

Texas Instruments, *Electromagnetic Emission from Logic Circuits*, Texas Instruments Inc., AR-SZZA007, November 1999.

Trobough, D., "Trends in PCB Materials," *Printed Circuit Design*, June 2001.

Williams, T., *EMC for Product Designers*, Oxford: Butterworth–Heinemann Ltd, 1992.

Yeager, J., and M. A. Hrusch-Tupta, *Low Level Measurements Precision DC Current, Voltage and Resistance Measurements*, 5th Edition, Cleveland: Keithley Instruments Inc, 1998.

Web Sites

http://www.qsl.net/wb6tpu/si_documents/docs.html

Application notes
http://bwcecom.belden.com/college/college.htm
http://www.avxcorp.com/
http://www.national.com/apnotes/apnotes_all_1.html
http://www.analog.com/techsupt/application_notes/application_notes.html
http://www.analog.com/publications/magazines/Dialogue/
http://www.fairchildsemi.com/apnotes/ap_notes_all.html
http://www.ti.com/sc/docs/apps/logic/buffers_and_driversapp.html
http://www.ti.com/sc/docs/apps/analog/power_managementapp.html
http://www.ti.com/sc/docs/apps/analog/amplifiers_and_comparatorsapp.html
http://www.onsemi.com/pub/prod/0,1824,productsm_Literature,00.html

MIL-STD docs can be found at:
http://stinet.dtic.mil/str/dodiss4_fields.html

13 LENSES, DISHES, AND ANTENNA ARRAYS

Lenses, dishes, and antenna arrays are all devices that can be used to concentrate and focus electromagnetic radiation. Furthermore, because these devices can isolate radiation from a specific direction, they can be used to create images or photographs.

REFLECTING DISHES

Dishes used for microwave transmission and reception are a good example to start with. Dishes are often used to receive microwave transmissions from satellites, such as those used for satellite television reception. The main purpose of the dish is to concentrate the microwaves, thus increasing the signal strength. The dish must be made of a reflective material; metal materials are used for radio and microwaves because metals reflect quite well. The dish is pointed in the direction of the source and it reflects the waves toward the center of the dish where an antenna is placed. For a dish to operate effectively, it must be at least several wavelengths in diameter, and the received signal strength is proportional to the area of the aperture. In practice, reflector dishes are usually at least 10λ in diameter. For this reason dishes are usually used only in the microwave region of the spectrum. If the dish is electrically small, the dish just scatters the radiation and its focusing properties are lost (diffraction is covered in Chapter 14). When any object is much larger than a wavelength in dimensions, the wave energy can be approximated as particles traveling in straight lines called rays. The approximation is referred to as the *geometrical optics approximation*. The theory of geometrical optics is used to described how lenses work. It can also be used to describe the behavior of dishes. *For apertures with diameter greater than about 60λ, diffraction can be mostly ignored, and the approximations of geometrical optics work quite well.* For smaller dishes, diffraction theory needs to be used, although geometrical optics will provide an approximation.

Figure 13.1 A reflecting dish showing the concentration of radiation.

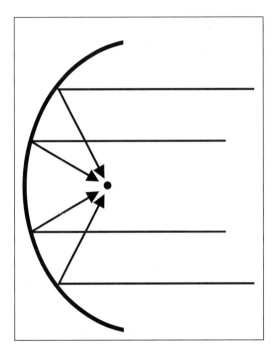

Figure 13.1 shows how radiation rays incident from broadside to the dish are concentrated at the antenna. This occurs because of the shape of the dish. The parabolic shape allows for all perpendicular rays to be reflected to the center, and the travel time for all rays to reach the center is equal. Thus, the rays reach the antenna in phase, and add constructively. Rays that arrive from other angles do not get concentrated at the antenna. Furthermore, the rays that do reach the antenna from other angles are not in phase, so they do not add constructively. Consequently the dish has a receiving/transmitting pattern with a thin lobe in the direction broadside to the antenna.

Figure 13.2 shows the receiving pattern of dishes as a function of electrical diameter. As you can see from this figure, larger dishes collect more radiation and serve to focus the radiation more narrowly. The width of the center lobe in each pattern contains about 90% of the total power. The center lobe width, $\Delta\theta$, can be calculated from

$$\Delta\theta = \sin^{-1}\left[\frac{\lambda}{2a}\right] \approx \frac{\lambda}{2a}$$

Figure 13.2 Receiving pattern for apertures (dishes, lenses, etc.) of various sizes. Note that the x-axis scale is not the same for the figures.

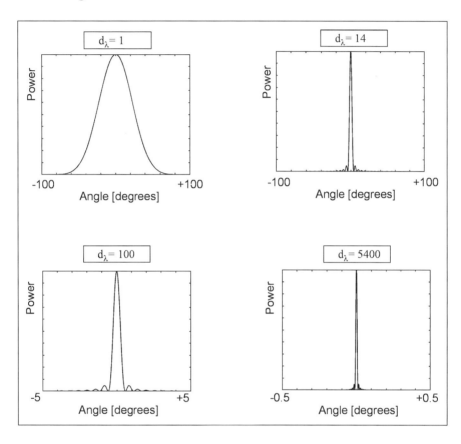

with the approximation being valid when the angular width, $\Delta\theta$, is small.

The 18 inch satellite dish used for satellite TV reception is a familiar example. The frequency of reception is centered at 12.45 GHz (Ku-band microwave), making this dish about 19 wavelengths in diameter. The rays are focused to a waveguide feedhorn antenna (a flared piece of waveguide). This Ku-band signal is amplified, down-converted to a signal that is centered at 1.2 GHz, and then output on 75 ohm coax that leads indoors to the receiver. The feedhorn and down-converter are integrated in a single package called a low-noise block-down (LNB) converter. In contrast, to create an equivalent dish in the FM radio band would require a dish diameter of about 60 meters. A dish

Figure 13.3 A reflecting dish focused at a finite distance.

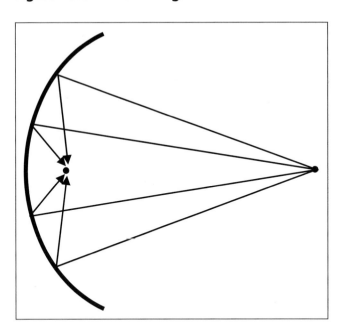

this size would take up the area of about half of a football field. A dish for the VHF TV band would require a comparable size dish. Now you can understand why you can't buy a dish antenna for receiving TV or radio signals—it would be too big to be practical. However, dishes of this size and larger have been built for the purposes of radio astronomy. For example, the radio dish at Arecibo, Puerto Rico, is 300 m in diameter. This dish has a movable antenna feed. By moving the antenna feed, the direction of focus can be changed. The dish can thus be used to scan across the sky.

The parabolic shape is useful for concentrating rays that are parallel. Parallel rays imply that the source is very far away, and the radiation takes the form of plane waves. In the terminology of cameras, the parabolic dish is focused at infinity. A reflecting dish can also focus rays that emanate from a source a finite distance from the dish by placing the feed antenna further from the dish. Figure 13.3 shows a dish that focuses rays from a finite distance. At a finite distance, the source rays form a radial pattern. These concepts will become more clear after you learn about lenses.

Microwaves and radio waves are not the only applications of reflecting dishes. The best optical telescopes actually use reflecting mirrored dishes instead of lenses. Very large mirrored dishes can be more easily and precisely manufactured than optical lenses of the same size. In an optical dish, a photographic plate, electronic camera, or other imaging device (e.g., the human eye) is used in place of an antenna.

LENSES

Instead of using a reflecting dish to focus rays of radiation, a lens can be used. Lenses are made from materials that are transparent (low loss dielectric) and which have a higher dielectric constant than air. In optical terminology, the index of refraction (n) is used in place of the dielectric constant (ε_r):

$$n = \sqrt{\varepsilon_r}$$

The wave impedance of the material can be calculated from the index of refraction:

$$Z = \sqrt{\frac{\mu}{\varepsilon}} = \frac{\eta_0}{n} \approx \frac{377}{n}$$

You have already learned that when waves encounter an impedance boundary they are partially transmitted and partially reflected. There is another effect that occurs when the radiation is incident at an angle other than perpendicular. This effect is the *refraction* or bending of light electromagnetic radiation. Figure 13.4 shows that when light rays are incident on a material of higher index of refraction, the rays change direction.

When going out the other side, the rays are bent once again. Refraction is the basis for how a lens works. The lens is constructed with a curvature such that rays incident at different points are bent by a different amount. Consider a lens focused at infinity, with rays incident perpendicular to the lens, as shown in Figure 13.5. The surfaces of this lens have a parabolic shape.

The ray that hits the center is exactly perpendicular to the surface of the lens, and therefore is not bent at all. Rays that hit the lens at points away from the center form an angle with the surface that is different than 90 degrees. These rays will thus be refracted (bent). In fact, the incident angle and thus the refraction angle change in proportion to the distance from the center of the lens. Therefore, rays at the edge of the lens will be bent the most. The result is that all the rays are focused at

Figure 13.4 Refraction of a ray of light passing through a layer of glass.

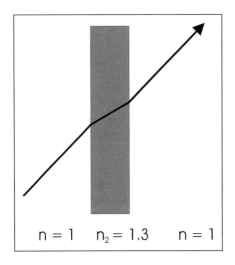

n = 1 n_2 = 1.3   n = 1

Figure 13.5 A lens focusing light from an infinite distance. The light converges at the focal length, f, of the lens.

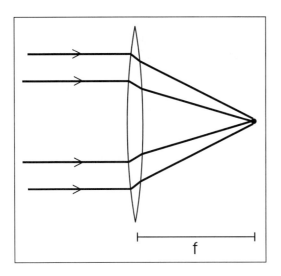

f

a single point behind the lens. This point is called the focal point of the lens, and is the location where the sensing element (antenna or photographic film, etc.) is placed. Another interesting point is that the waves that follow each ray have exactly the same path length, and therefore arrive synchronized in phase. The rays may have different lengths, but

IMAGING

the slow speed of light inside the lens ($v = c/n$) causes delay. For example, the ray that travels through the exact center is delayed the most because it travels through the thickest part of the lens. The rays that travel through the lens near the edge are delayed the least because they travel through the thinnest part of the lens. These rays, however, travel the longest distance through air. The physics works out such that the delay through air plus the delay through the lens is always the same for rays that arrive at the focal point.

The lens just described performs the same function as the parabolic reflecting dish. The receiving pattern of such a lens is identical to those of the reflecting dish shown in Figure 13.2. You could even make an 18 inch lens to use instead of the satellite TV dish. Instead of placing the antenna feed in front, the antenna feed would be placed in back of the lens, to collect the microwave radiation. Glass is a low-loss dielectric in the microwave region, with an index of refraction of about $n = 2.25$. The index of refraction is actually slightly higher at the microwave frequencies than at visible frequencies. The reflecting dish is used for microwaves because a dish of metal is much cheaper to manufacture than a glass lens. The dish is also much sturdier. Of course, making a lens at radio frequencies, such as the FM radio band, is also possible in theory, but obviously not viable due to the immense aperture size required.

IMAGING

Because the focal point stands behind the lens, it is very easy to use a lens for imaging. Suppose that you place a photographic plate in the plane of the focal point as shown by Figure 13.6. The plane where the plate is located is called the *image plane*.

Assume that the lens has a diameter of very many wavelengths. Not only will rays perpendicular to the lens focus to a point at the center of the image plane, but rays that arrive at slightly different angles will focus to points slightly offset from the center. In fact, for every angle of incidence of a plane wave, there is a corresponding point on the image plane where light from that direction will be focused. A photographic plate as well as the special receptor cells in our eyes can capture the light at all of the different image plane points simultaneously. Thus, both systems (the camera and the eye) can produce images. The imaging system shown in Figure 13.5, which is focused at infinity, can be used to image objects at great distances such as stars in the sky or mountains in the distance. The distance, *f*, from the lens to the focal plane is a function

Figure 13.6 A lens focused on an object (the gray arrow) produces an image at the image plane. The rays of light from the head of the arrow are shown.

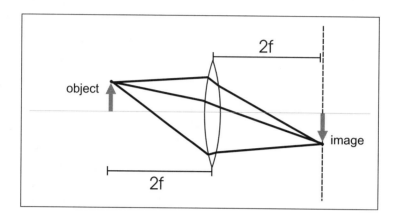

Figure 13.7 A lens focusing light from a finite distance.

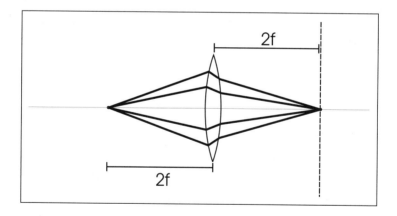

of the lens material index of refraction and of the shape of the front and back surfaces of the lens. In practice, lenses usually have a spherical surface instead of a parabolic surface because the spherical surface is easier to create. Some slight error occurs due to this shape. The focal length for a thin spherical lens is a function of the radius of curvature of the two spherical surfaces, front and back.

Imaging is not limited to objects at very large and approximately infinite distances. Any imaging system can be easily changed to focus on near objects (Figure 13.7) by increasing the distance from the lens to the imaging plane. This is how a camera is focused. The lens is moved in relation to the imaging plane where the film resides. The human eye uses a different method to focus. The muscles of the eye can be tensed to change the shape of the lens, thus changing the focal properties of the lens.

Imaging Resolution

All imaging devices are limited in the smallness and spacing of objects they can resolve. Objects smaller or more closely spaced than the device's resolution become blurred. There are two limiting factors of resolution. The first is the physical wavelength. Due to the phenomenon of diffraction, which is common to all wave types, you cannot image objects smaller than about a half-wavelength in dimension. For this same reason transmitted wave energy is greatly diminished when incident upon small apertures less than $1/2\lambda$ in both dimensions. The waves cannot easily pass through such small openings. The complement to a small aperture is a small object. When waves are incident upon a small disk, for instance, very little of the wave is reflected. The wave tends to diffract around the object instead. This property of waves defines the resolution limit for microscopes. Visible-light microscopes can only resolve objects down to about one half of the shortest visible wavelength. The shortest visible waves are in the violet color range and have a wavelength of about 380 nm. This same property limits the resolution of the photolithographic processes used in integrated circuits manufacturing, although in the lithographic process radiation in the ultraviolet can be used because the process is not limited by the properties of human vision.

The second resolution limit is the angular resolution limit. The angular resolution defines how closely an imaging device can resolve two objects in terms of angular separation. The angular resolution of an imaging system is dependent on the electrical diameter of the aperture. An aperture is in effect a window of the entire imaging signal. The resulting image is blurred in inverse proportion to the aperture's electrical dimension. An aperture with an electrical radius tending to infinity produces an image tending to infinite angular resolution.

Angular resolution is defined by the Rayleigh criterion. The Rayleigh criterion defines the minimal angular resolution in terms of the electrical size of the aperture (e.g., lens, dish, or antenna array):

$$\theta = \frac{1.22}{D_\lambda}$$

where D_λ is the aperture diameter given in wavelengths. Smaller wavelengths and larger apertures provide better angular resolution. For this reason cameras with larger lenses produce sharper images. This fact can be readily seen by examining Figure 13.2. Apertures that are electrically larger produce sharper focal lobes. The human eye has an angular resolution of about 0.01 degrees. The angular resolution is what limits our vision in practice. Large telescopes have much larger optical apertures and can therefore provide much higher resolution images of the sky. As I highlighted in Table 1.1, imaging becomes impractically large at radio frequencies. The eye has a diameter of about 5000 wavelengths. To create an imaging device of equal electrical diameter requires an aperture of diameter of 1600 m at 100 MHz. In Chapter 1, I somewhat arbitrarily defined an aperture of minimal imaging as 14 wavelengths. Using the Rayleigh criterion, such an aperture can resolve angular separations of 5 degrees.* Microwaves can be used for imaging, but the resolution of the objects it can measure is much less than that in the optical range. Microwave imaging has several applications, including two-dimensional radar imaging (pictures) for aircraft identification. Traditional one-dimensional radar just produces a single dot on a radar screen, whereas two-dimensional radar produces an image of aircraft similar to a low-resolution black and white photo.

ELECTRONIC IMAGING AND ANTENNA ARRAYS

Imaging systems such as the eye and electronic video cameras create images that are sampled in space. In other words, there is an array of light-sensitive elements at the imaging plane, and each element produces its own electronic signal proportional to the light intensity received at its location on the imaging surface. If the array is very finely grained, the image resolution is unaffected. If the array is not finely grained, a pixilated effect occurs.

Now, what if the "middle man" (i.e., the lens), is removed? Could electronics somehow process the radiation received by the sensor elements as a lens would, such that an image could be created? The answer is yes! Although the technology does not exist for such processing to be accomplished at visible frequencies, electronic imaging can be accomplished readily at radio and microwave frequencies. The key factor for electronic imaging is the ability to process all the information in the wave. Visible light sensing elements can only produce a signal proportional to the time-averaged amplitude (intensity) of the received light.

---

*The moon provides a gauge for understanding angular size. As seen from Earth, the moon subtends 1/2 degree.

ELECTRONIC IMAGING AND ANTENNA ARRAYS

Figure 13.8 A two-dimensional array of dipole antennas forms a rectangular aperture, sampled at discrete points in space.

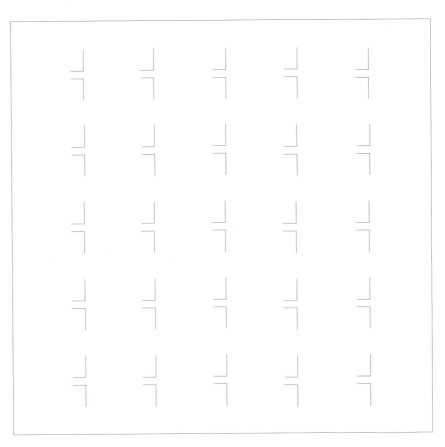

All phase information is lost. In contrast, radio antennas capture the actual wave, providing instantaneous amplitude and phase of the radiation received at that location.

An antenna array is a group of antennas spaced such that they span an electrically large aperture. There are many variations on antenna array geometry. Some are very simple, composed of only two or three antennas, while others comprise hundreds of antennas. I will only discuss the basic premises of antenna arrays. Suppose you want to create an aperture for imaging at radio or microwave frequencies, but you don't want to use a dish. Instead you can create an electronic lens, by placing numerous antennas evenly across an aperture, as shown one-dimensionally in Figure 13.8.

In effect, this is a sampling of a continuous aperture at discrete locations, the spatial analogue to the time domain sampling performed in digital audio systems. Time domain sampling in audio systems puts a limit on the minimum time resolution (or equivalently, it puts a limit on the maximum frequency). The spatial case is somewhat more complex, and I will not cover the details. The processing to produce an image is quite simple—delay and sum. The electronics need to perform the delay that physically occurs in a lens or reflecting dish, such that all the rays are added together in phase. For example, to focus far-field radiation received from the direction broadside to the array, the antenna signals are just added together. To focus far-field radiation received from an angle, θ, the signals are each delayed by amount,

$$\tau = \frac{x \times \cos\theta}{c}$$

to recreate the wavefront (places of equal phase) as shown in Figure 13.8. In this formula, x is the distance along the array, θ is the angle in the direction of the source, and c is the speed of light. This technique allows for focusing at infinity. For focusing objects at a finite distance, each signal must be delayed to recreate the spherical wavefront of close objects. For example, to focus the antenna array at a finite distance, r, broadside to the array, you apply a delay,

$$\tau = \frac{\sqrt{r^2 + x^2}}{c} - r \approx \frac{x^2}{2rc}$$

to each element. In this formula, x is the distance from the center of the array and r is the distance to the source from the center of the array. The delay can be implemented via LC filters or via computer programs in a microprocessor (assuming that the signals are digitized).

The application of antenna arrays is not limited to imaging. Antenna arrays can be used to focus signal reception from a specific direction, just like a reflecting dish. If the signals are narrowband, and therefore approximately single frequency, a simple phase shifter can be used instead of a delay function. Most communication signals fall in this category and are referred to as *coherent* sources. The process of using electronics to focus an antenna array is called *beamforming*. The beauty of the antenna array is that its beam direction can be changed electronically, in contrast to the reflecting dish, which must be physically redirected using motors. Another advantage of the antenna array is that it can simultaneously scan signals from different directions by operating multiple microprocessing algorithms or multiple circuits in parallel. The reflecting dish can only scan one direction at a time. The fact that antenna array signals are processed via electronics allows for adaptive antennas and so-called "smart antennas" to be designed. The last benefit of antenna arrays is that they can be used to create immense apertures, covering thousands of meters. All that is needed is a set of well-spaced antennas.

OPTICS AND NATURE

It is an interesting fact of nature that the visible portion of the spectrum is really the only viable part of the spectrum for animal imaging. (Portions of the near infrared and near ultraviolet, which surround the optical range, can also be used and are used by some animals. For example, bees and many birds can see into the ultraviolet, and turtles can see into the infrared.) Radio and microwaves are too large to resolve objects such as small animals and insects. Water contains atomic and molecular resonances through most of the infrared, making it opaque at these frequencies. At radio frequencies, water becomes transparent again, but radio wavelengths require gigantic apertures. In addition, due to the large wavelength, radio wave imaging devices can resolve only large objects. Even if 1 GHz were used, the aperture would need to be at least several meters in diameter, and could only resolve objects of 15 cm or larger—not very useful for any animal on Earth. Water also becomes opaque (lossy) in the higher ultraviolet range. X-rays are too scarce and too damaging to be used for animal imaging. Finally, the sun's radiation peaks near the infrared/visible border and is filtered by water and gases in the atmosphere, making visible light the most plentiful region of the natural spectrum received on Earth. Animal vision is optimal in many ways.

BIBLIOGRAPHY AND SUGGESTIONS FOR FURTHER READING

Andrews, C. L., *Optics of the Electromagnetic Spectrum*, Englewood Cliffs, NJ: Prentice-Hall, Inc., 1960.

Born, M., and E. Wolf, *Principles of Optics*, 7th Edition, Cambridge, UK: Cambridge University Press, 1999.

Elliot, R. S., *Antenna Theory and Design*, Englewood Cliffs, NJ: Prentice-Hall, 1981.

Heald, M., and J. Marion, *Classical Electromagnetic Radiation*, 3rd Edition, Fort Worth, Tex.: Saunders College Publishing, 1980.

Hecht, E., and K. Guardino, *Optics*, 3rd edition, Reading, Mass.: Addison-Wesley, 1997.

Johnson, D. H., and D. E. Dudgeon, *Array Signal Processing: Concepts and Techniques,* Englewood Cliffs, NJ: Prentice Hall, 1993.

Justice, J. H., N. L. Owsley, J. L. Yen, A. C. Kak, *Array Signal Processing,* Englewood Cliffs, NJ: Prentice-Hall, Inc., 1985.

Kraus, J. D., *Antennas,* 2nd Edition, Boston: McGraw-Hill, 1988.

Pedrotti, F. L., and L. S. Pedrotti, *Introduction to Optics,* 2nd Edition, Upper Saddle River, NJ: Prentice Hall, 1993.

Yu, F. T. S., *Optical Information Processing,* Malabar, Florida: Robert E. Kreiger Publishing Company, 1990.

14 DIFFRACTION

Light, and all electromagnetic waves in general, does not necessary follow a straight path; electromagnetic waves can bend around objects. This property is common to all waves, not just electromagnetic waves. For example, consider the propagation of water waves in Figure 14.1.

The waves are traveling from left to right through open water. Upon encountering a wall with a small opening, the waves diffract. That is, on the right side of the wall, the waves propagate from the aperture in a circular manner. This behavior is in direct contrast to how particles progress through an aperture, as shown in Figure 14.2.

As with the waves, only the portion of the incoming particles that arrive at the opening can make it through to the right-hand side. However, in contrast, the particles that do make it through the aperture continue to travel in straight lines.

In general terms, diffraction is the aspect of light that cannot be described by geometrical optics. In other words, diffraction effects are those effects that cannot be explained from the physics of particles. Having studied the physics of lenses and geometrical optics, Newton was convinced that light consisted of particles that traveled in straight lines. Although other scientists of his time believed that light was a wave, it wasn't until many experiments of light diffraction were conducted that science as a whole was convinced of the wave nature of light. Ironically, quantum physics brought back the particle nature of light. The modern view is that light can somehow act as both a particle and a wave.

DIFFRACTION AND ELECTRICAL SIZE

Diffraction is very dependent on the electrical size of the objects involved. With objects having dimensions below a wavelength, geometrical optics cannot be used at all. Diffraction effects dominate when waves interact with these small objects. On the other hand, the interaction of waves with objects whose dimensions are greater than about

Figure 14.1 Water waves incident upon a wall with an aperture (opening).

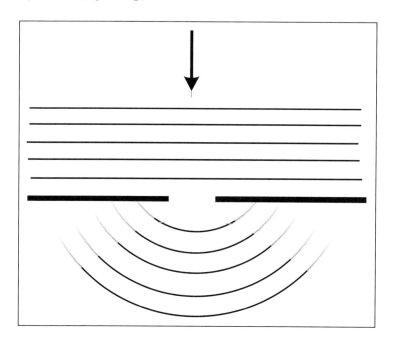

Figure 14.2 Particles incident upon a wall with an aperture.

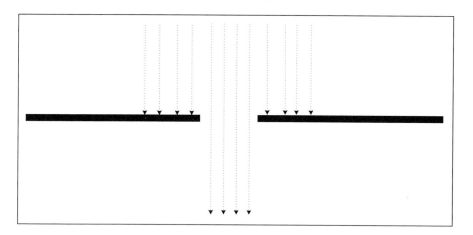

60λ is governed mostly by geometrical optics, and diffraction effects usually go unnoticed. Diffraction accounts for much of the difference between light wave propagation and radio wave propagation. For example, imagine you and a friend standing on opposite sides of a perfectly absorbing, 10 foot–high wall. If your friend shines a light at the

Figure 14.3 Demonstration of Huygens' principle. Each point of the initial wavefront creates its own spherical wavelet. The next wavefront (*dotted line*) is created from the outline of the individual wavelets.

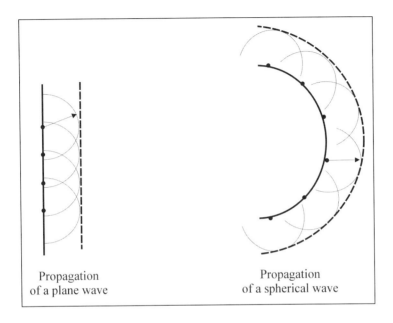

wall, none of the light will make it to you. However, if he sends a message with a radio wave at say 1 MHz, the wave will easily diffract around the wall and reach you.

HUYGENS' PRINCIPLE

One of the best ways to picture how light waves propagate is through the application of Huygens' principle. Starting with an arbitrary wavefront, you assume that each point along the wavefront is a point source for a new propagating wavelet. By taking the envelope of all the wavelets from each point, you produce a second wavefront. This process can be continued again and again to produce successive wavefronts. Figure 14.3 illustrates this process. This process is only an approximation. It does not account for the effects of wavelength. Fresnel produced an exact method of wave construction with a slight modification to Huygens' principle. The Huygens-Fresnel principle states that you sum the wavelets from each point, including both amplitude and phase, on the wavefront. In other words, you allow each wavelet to interfere.

Figure 14.4 Illustration of Babinet's principle. The sum of the light patterns of an aperture and an object of equal size and shape is equal to the pattern without any barriers.

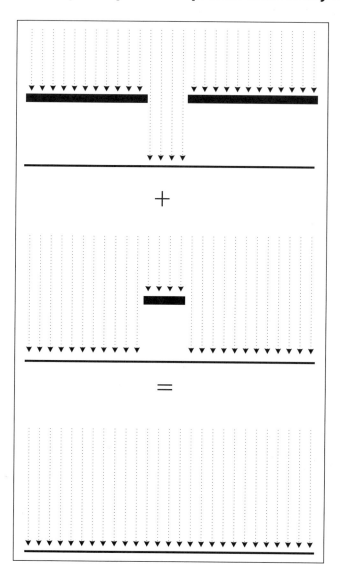

BABINET'S PRINCIPLE

Consider the three situations depicted in Figure 14.4. In each case, a plane wave source is incident upon an imaging plane. In the first case, there is no intervening object, and the light intensity is constant across the image plane. In the second case, an object is placed before the imaging plane and a shadow is cast on the imaging plane. In the third case, a screen with a single aperture is placed in front of the imaging plane. The aperture has the same dimensions as the object in case 2. The aperture is also positioned at the same place as the object from case 2. A small region of light is cast onto the imaging plane, with the rest of the plane in shadow. Babinet's principle states that if you sum the wave amplitude (not the power or intensity, which is related to the square of the amplitude) at the imaging plane for cases 2 and 3, the result is the image from case 1. This principle holds in general for any wave source, and any complementary set of screens.

In 1870, Lord Rayleigh derived the law for scattering of light from objects much smaller than a wavelength. The law states that the scattered light is proportional to the area of the object multiplied by the fourth power of the electrical dimension of the object. This result is the complement of the transmission of light through small apertures, which I discussed in Chapter 9. In 1944, Nobel Prize–winning physicist Hans Bethe derived the theory of diffraction for small apertures. The equivalent result of the two theories demonstrates the success of Babinet's principle.

FRAUNHOFER AND FRESNEL DIFFRACTION

Diffraction is a radiation phenomenon. Fraunhofer diffraction occurs when the wave source and observer are far from the aperture or scattering object. Fresnel diffraction occurs when the source and/or observer are close to the aperture or scattering object. Figure 14.5 shows Fresnel diffraction patterns for apertures of various cross-sections. Figure 14.6 shows Fresnel diffraction patterns for opaque (perfectly absorbing) objects with different size cross-sections. In other words, Figure 14.6 shows the Babinet complements to those objects in Figure 14.5. In each case, the source frequency is 100 MHz ($\lambda \times 3$ m), and the source antenna is assumed to be very far away from the objects. The electric field is measured at a distance of 10 m from the object. For electrically large objects, the shadow is in proportion to the object. The shadows cast by ordinary everyday objects under illumination by visible light fall into this category. For electrically small objects, the shadow becomes almost impossible to discern. The shadows cast by ordinary everyday objects when illuminated by radio waves fall into this category.

Figure 14.7 shows a close-up view of a shadow edge. The shadow does not immediately change from dark to light, but goes through a transition period on the order of several wavelengths.

Figure 14.5 Simulated diffraction patterns for apertures of various widths, with incident radiation of 100 MHz (3 m wavelength). Patterns "measured" 10 m from aperture. The figures graph the intensity of the radiation pattern.

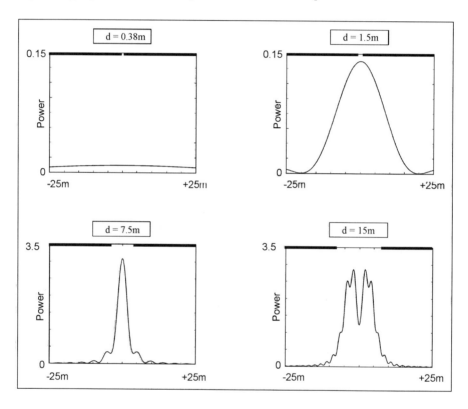

Diffraction patterns depend on both the ratio of aperture to wavelength (electrical size of the aperture) and the ratio of the aperture to distance from aperture to the imaging plane. Notably, the pattern spreads as the relative distance is increased from aperture to imaging plane. This result is commonly observed whenever you use a flashlight. The flashlight beam broadens as it travels away from the flashlight.

RADIO PROPAGATION

The propagation of radio waves displays many diffraction effects. Radio waves do not create crisp shadows because the wavelengths are much larger than the objects encountered. Diffraction also allows radio waves

Figure 14.6 Simulated diffraction patterns for objects of various widths, with incident radiation of 100 MHz (3 m wavelength). Patterns "measured" 10 m from object. The figures graph the intensity of the radiation pattern.

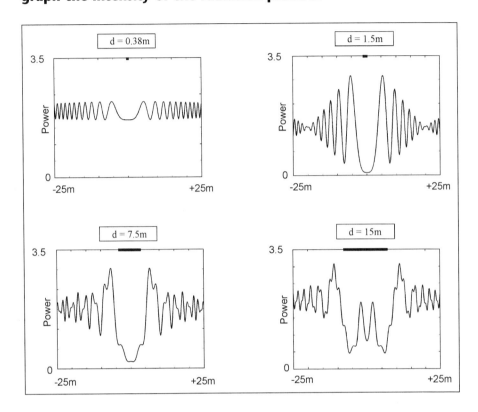

to travel over the horizon. The waves are diffracted by the surface of the Earth.

CONTINUOUS MEDIA

Waves propagate around electrically small objects without any crisp shadows. If many small objects are evenly distributed in the path of a wave, the effect can be approximated by considering the medium as continuous, with effective dielectric properties. For radio waves, rain and tree leaves are examples of small objects that can be approximated by assuming a continuous medium with slightly different electrical properties than those of air.

Figure 14.7 Simulated diffraction patterns at the edge of a long object, with incident radiation of 100 MHz (3 m wavelength). Pattern "measured" 10 m from object. The figures graph the intensity of the radiation pattern.

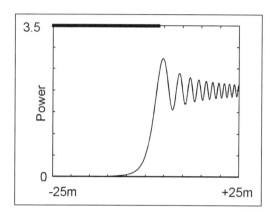

In fact, this approximation is used throughout electrical engineering. No material is truly continuous. Rather, materials are composed of individual particles—electrons, molecules, ions, and atoms. For radio waves, infrared, visible, and even ultraviolet radiation materials can be considered to be continuous media with properties of conductivity, permittivity, and permeability. Only when the radiation wavelength approaches the size of the molecules and atoms and their spacing does their discrete nature become apparent. High-frequency X-rays have wavelengths small enough to be affected by individual atoms in materials, and can be used to image the structure of crystalline materials.

BIBLIOGRAPHY AND SUGGESTIONS FOR FURTHER READING

Andrews, C. L., *Optics of the Electromagnetic Spectrum,* Upper Saddle River, NJ: Prentice-Hall, Inc., 1960.

Boithias, L., *Radio Wave Propagation,* Boston: McGraw-Hill, 1987.

Born, M., and E. Wolf, *Principles of Optics,* 7th Edition, Cambridge, UK: Cambridge University Press, 1999.

Heald, M., and J. Marion, *Classical Electromagnetic Radiation,* 3rd Edition, Fort Worth, Texas: Saunders College Publishing, 1980.

Hecht, E., and K. Guardino, *Optics,* 3rd Edition, Reading, Mass.: Addison-Wesley, 1997.

Jackson, J. D., *Classical Electrodynamics*, 2nd Edition, New York: John Wiley & Sons, 1975.

Pedrotti, F. L., and L. S. Pedrotti, *Introduction to Optics*, 2nd Edition, Upper Saddle River, NJ: Prentice Hall, 1993.

Yu, F. T. S., *Optical Information Processing,* Malabar, Florida: Robert E. Kreiger Publishing Company, 1990.

15 FREQUENCY DEPENDENCE OF MATERIALS, THERMAL RADIATION, AND NOISE

This the final chapter covers some miscellaneous but interesting topics that relate to electromagnetics. An often overlooked aspect of electromagnetics is the fact that all materials possess varying electromagnetic properties in different regions of the electromagnetic spectrum. I also cover the basics of thermal radiation, which is the type of electromagnetic radiation responsible for radiative heat transfer, and thermal circuit noise. Both thermal radiation and thermal circuit noise are caused by the random motion of charged particles, most notably electrons, inside a material.

FREQUENCY DEPENDENCE OF MATERIALS

In electronics you spend your time with materials that have (approximately) constant properties in relation to frequency. However, no material retains the same response over all frequencies. Many materials are relatively constant at electronic frequencies, but almost all materials undergo property changes in the infrared and/or visible bands. We are quite fortunate that many materials go through changes in the visible band, because the variations provide the many different colors that we see and utilize to identify objects.

Conductors and Dielectrics

There are two basic types of materials, conductors and dielectrics. Conductors are materials that support a current at DC. As such, they have large concentrations of mobile charge. Most good conductors are metals, but ionic solutions such as salt water are one example of nonmetal conductors. The other major type of material is the dielectric or insulator. Dielectrics are poor conductors at DC. There are of course materials that fall into the gray area in between the two types, but I will focus on the distinctions to keep the topic clear.

Metals

Because metals are the most commonly used conductors, I will focus on the properties of metallic conductors. At DC and radio frequencies, the electrons inside the metal can be approximated as a classical gas of charged particles bounded by the surface of the metal. You can picture this model as tiny charged balls that are constantly in motion bouncing off one another. When an electric field is applied, the electrons experience a net drift in the direction of the field. Because there is such a high density of electrons, no electron can travel very far before it collides with another electron or with an atom in the lattice of the metal. The applied electric field causes more motion in the electron gas. Due to collisions, some of the applied energy is lost to the random electron motion and atomic vibrations, which is simply the microscopic manifestation of heat. The energy lost to heat is proportional to the resistivity of the metal multiplied by the current squared. A superconductor exhibits no electron collisions, no heat loss, and no resistivity.

If an AC signal is applied to the metal, the effect is basically the same, except that the electrons are sloshing back and forth in synchronization with the changes in the applied field. This describes metallic behavior from DC through microwaves. However, in the infrared the wave sloshing becomes so rapid that the wave can change directions before some of the electrons experience a collision. The frequency at which this starts to occur is called the *relaxation* or *damping frequency* of the metal. For copper this frequency occurs at about 4 THz, as can be seen in Figure 15.1. At much higher frequencies, such as the visible range, the electrons move with the wave without having any more collisions than they would without the wave. In other words, there is no heat produced by the electric field. A metal at these frequencies is said to behave as a collision-less plasma. Metals are reflective both below and above the relaxation frequency. This characteristic is pointed out every time we look in the mirror. Assuming that electronics progresses into THz, you may wonder how the change in metal properties will affect the use of metals for transmission. The physics involved seems to allow transmission lines to be created well above the relaxation frequency.

However, there is a major problem to creating transmission lines beyond the terahertz regime: scaling laws. Scaling laws occur throughout nature. For example, a typical human can only carry objects that are about the same weight as his or her body, whereas an ant can carry objects that are 100 times its own weight. The scaling laws of

Figure 15.1 Calculated electrical properties of copper as a function of frequency. The conductivity is constant up to the relaxation frequency (~4THz), where it starts to drop. The reflectivity is close to 1 until the visible range. The absorption increases up to the near infrared, where it levels off. In the ultraviolet, the absorption rapidly decreases.

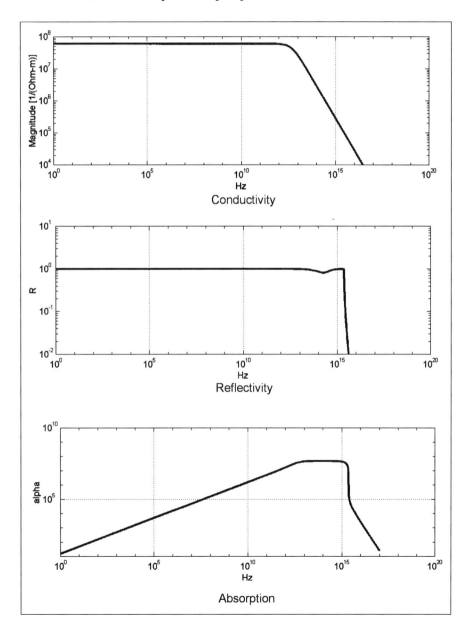

electromagnetics are such that signal attenuation on transmission lines is inversely proportional to the spacing of the conductors. Thus, the infrared signal attenuation becomes too great for transmission lines to be practical.

Metals exhibit a second change in properties that occurs above the visible range. This transition frequency is referred to as the *plasma frequency*. Above the plasma frequency, a metal becomes transparent! In fact most metals become transparent in the ultraviolet range. This same phenomenon occurs in the ionosphere. The ionosphere consists of a gas of ions and free electrons, forming a plasma. At low radio frequencies, the ionosphere reflects radiation, allowing skip-mode propagation and global waveguide effects. In contrast, the microwave range is above the plasma frequency of the ionosphere. The ionosphere is transparent to microwaves, allowing for communication between the Earth and orbiting satellites. The plasma frequency of the ionosphere occurs in the VHF band.

Dielectrics

Dielectrics are materials that do not allow DC conduction. This property implies that dielectrics do not have charges that are free to move around. This does not imply that the dielectrics do not respond to electric fields. Any material with a dielectric constant above that of a vacuum has some response to electric fields. On the molecular or atomic scale, charge is displaced via rotation or separation. In water, the major reaction is that of rotation of the H_2O molecule, which is inherently polarized. Another type of reaction is ionic polarization; the bond between atoms that make up a molecule is stretched and relaxed when placed in an electric field. The third major type of polarization is electronic polarization. Electronic polarization is the stretching of the electron cloud (orbital) around the nucleus. In all of these phenomena, the polarization occurs in a finite, non-zero amount of time. That is, it takes a certain amount of time for the polarization to occur. Think of a pendulum—it takes time for it to move back and forth. The pendulum also has a resonant frequency corresponding to how quickly it can move and then return to the same position. When the frequency of the applied field is low compared to the resonant frequency, the polarizing charge can perfectly track the changing electric field. When the frequency becomes greater than the resonant frequency, the charge can no longer track the applied field quickly enough. Therefore, above this frequency polarization does not occur and the dielectric constant drops in magnitude.

FREQUENCY DEPENDENCE OF MATERIALS 335

A side-effect of the polarization of charges is that of heat loss. This process is friction on a microscopic scale. When the charges are polarized in a dielectric, some energy is lost to random vibrations in the material. In dielectrics, the heat loss peaks at the polarization resonant frequencies.

The summary just given is very simplistic. All real materials exhibit combinations of many different effects. Any slight impurities in a material such as water or glass will also contribute. Furthermore, in the infrared and above, electrons can be excited into higher quantum orbitals, thus absorbing radiation at specific frequencies. Any irregularities or roughness that is electrically large will also contribute to scattering and absorption. Most everyday objects are compounds of simple and complex molecules, producing a very varied response in the visible range.

Water

Water is an example of a dielectric. Even this simple molecule has a varied frequency response. Near DC and at low radio frequencies, water is a good insulator. At these low frequencies water is transparent to radiation and is low loss. As frequency increases, so does loss. Water is quite lossy even in the microwave and is extremely lossy in the infrared, making it opaque to radiation. The loss drops dramatically in the near infrared, leading to the well-known transparent nature of water at visible frequencies. It becomes opaque again in the ultraviolet. Any dissolved chemicals such as salt dramatically alter the properties of water, making it a DC conductor and making it much lossier in the radio frequencies. This fact makes radio communication with submarines very difficult. Very low frequencies (~10 kHz) are used to communicate with submarines. The electrical properties of water are shown in Figure 15.2.

Glass

Glass is a dielectric that is quite similar to water in many respects. As exhibited in Figure 15.3, glass is transparent at radio and visible frequencies and is opaque through much of the infrared and ultraviolet regions. The opaqueness of glass in the ultraviolet explains why you don't get sunburn from sunlight that passes through glass. Another useful property of glass is its opaqueness in the infrared. This property is exploited in greenhouses. Visible and some near infrared radiation from the sun passes through the glass and is absorbed by the plants within the greenhouse. The plants, being at a much lower temperature than the sun, reradiate their heat in the far infrared. Glass is very lossy in the far infrared, preventing the radiation from escaping.

Figure 15.2 Electrical properties of water as a function of frequency. Dotted lines mark the visible region of the spectrum. Data compiled from D.J. Segelstein, "The complex refractive index of water," M.S. Thesis, University of Missouri-Kansas City, (1981). Data available online at http://omlc.ogi.edu/spectra/index/html.

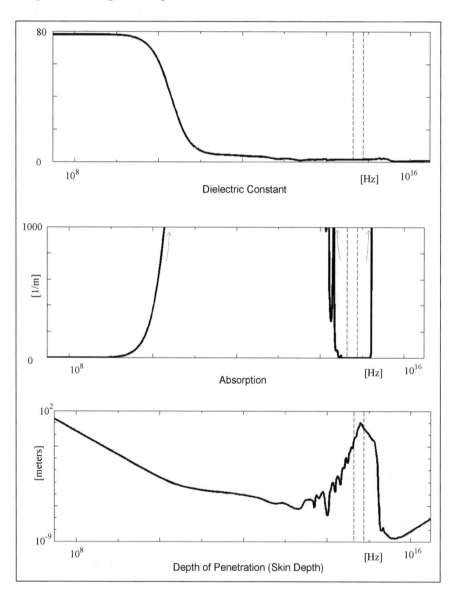

Figure 15.3 Electrical properties of glass (fused quartz) as a function of frequency. Inset figure shows a closeup of the visible region. Data compiled from T. Henning and H. Mutschke, "Low-temperature infrared properties of cosmic dust analogues," *Astronomy and Astrophysics*, vol 327, pp. 743–754. (http://urania.astro. Spdu,ru/JPDOC/1-entry.html) and Crystran IR-grade fused data sheet (http://www.crystran.co.uk/).

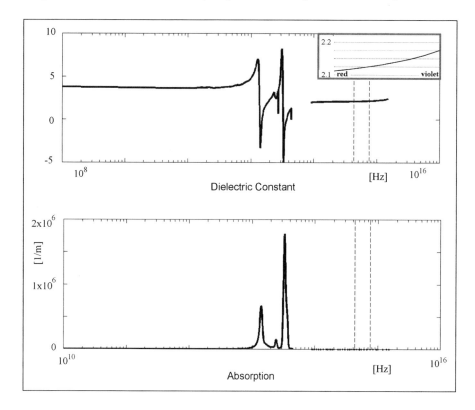

Glass prisms take advantage of the slight change in glass permittivity in the visible region. For example, flint glass has an index of refraction of about 1.71 in the red frequencies and about 1.73 in the blue frequencies. When white light is incident on a prism, the different colors are each refracted by slightly different angles, producing a rainbow effect. Rainbows seen in the sky are due to the same property in water droplets.

There are many types of glass, and the dielectric constant will vary between glass types. At electronic frequencies (microwave and below), the dielectric constant can range from about $\varepsilon_r = 3.75$ to $\varepsilon_r = 7.0$. In the visible range, the dielectric constant can range from about $\varepsilon_r = 1.5$ to $\varepsilon_r = 3.0$.

Figure 15.4 Penetration depth of waves into the Earth's surface: (A) sea water, (B) humid soil, (C) fresh water, (D) moderately dry soil, (E) very dry soil. Reproduced from Boithas, L., *Radio Wave Propagation*, Boston: McGraw-Hill, 1987, p. 77, with permission of Kogan-Page, Ltd.

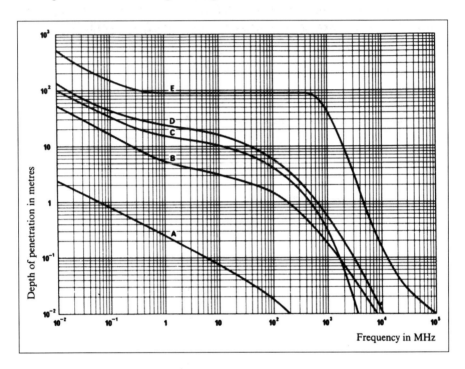

Ground

The Earth has very different properties across the electromagnetic spectrum. The Earth is a decent conductor in most locations because the soil contains water and dissolved salts. However, the conductive properties of soil change throughout the radio bands, and are different depending on the wave polarization (horizontal or vertical) and angle of incidence. The frequency-dependent nature of soil conductivity has dramatic effects on radio propagation at different bands and is very important to the design of antenna broadcast and transmission systems. The frequency dependence of Earth is illustrated in Figures 15.4 through 15.6.

HEAT RADIATION

Thermal energy is the random kinetic energy of the microscopic particles (electrons, atoms, ions, molecules) that constitute matter. One of

Figure 15.5 Permittivity and conductivity of the Earth's surface: (A) sea water, (B) humid soil, (C) fresh water, (D) moderately dry soil, (E) very dry soil, (F) pure water, (G) ice. Reproduced from Boithas, L., *Radio Wave Propagation*, Boston: McGraw-Hill, 1987, p. 53, with permission of Kogan-Page, Ltd.

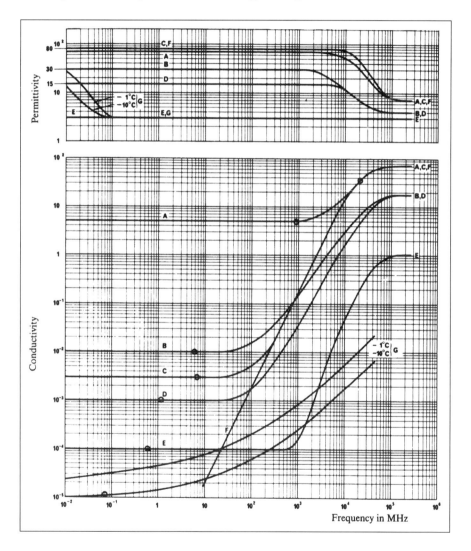

the crowning achievements of 19th-century physics was the development of the science of thermodynamics. Possibly the most fascinating aspect of thermodynamics is the kinetic theory of temperature. Temperature is simply a measure of the average speed of the particles in an object. To be precise, temperature is proportional to the mean-square

Figure 15.6 Soil reflectivity (moderately dry soil): dashed line represents horizontal polarization, solid line represents vertical polarization. Reproduced from Boithas, L., *Radio Wave Propagation*, Boston: McGraw-Hill, 1987, p. 55, with permission of Kogan-Page, Ltd.

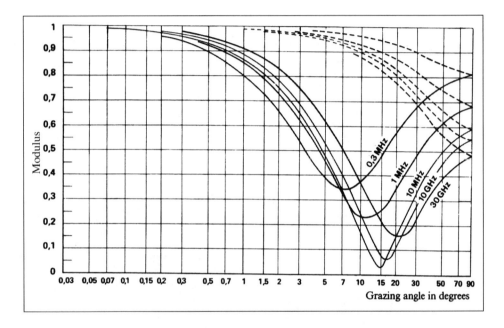

velocity of the particles. For example, in a gas molecules are free to move at will and collide with one another at random. When the temperature of a gas increases, the molecules move faster and collisions occur more often. In a solid, the molecules, atoms, and/or ions that constitute the solid are held in place. The particles in solids are not free to move about, but they do vibrate. The vibrational motion of these particles is how thermal energy is stored and conducted in solids.

In Chapter 5, I stated that any charged particle that is accelerated (or decelerated) will radiate energy in proportion to the magnitude of the acceleration. In electronic applications, the electrons of a metal typically are accelerated in a sinusoidally by an applied AC field, producing sinusoidal radiation. This is a special, manmade form of radiation. Because the electrons are always in random motion due to heat contained in the metal, all metals also exhibit a natural radiation—that of heat radiation. In Chapter 2, I discussed how electrons in metals are free to move and

Figure 15.7 Thermal radiation spectrum of black body at 37°C (100°F). Note that the radiation in the visible band is about 23 orders of magnitude less than the peak radiation in the infrared.

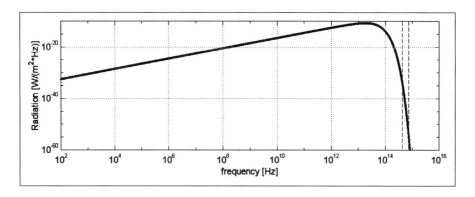

are in constant motion. The electrons can be approximated as a gas of charged particles. They are ever moving, colliding, and changing direction. The impulses caused by the collisions imply that electrons are experiencing acceleration and deceleration. In other words, the electrons are not moving at a constant velocity. The random movement of the electrons produces a random output of heat radiation. Much of this energy is absorbed by other electrons in the metal, but some is also radiated beyond the metal. Approximating a resistor as a perfect black body (i.e., a perfect radiator/absorber), you can calculate the spectrum of its radiation. Let's assume that the resistor is at about 37°C (100°F).

Figure 15.7 shows the spectrum of heat radiation from an object, such as a resistor or a human, at 100°F. Note that the radiation peaks at about 20 THz and drops by about 23 orders of magnitude (640 dB) in the visible band. This fact explains why we can't see the thermal radiation of room temperature (or near room temperature) objects. On the other hand, at 1 THz, the radiation is only about 40 dB down from the peak. Radiation noise is quite strong in the millimeter waves and infrared.

The radiation spectrum can be calculated using Planck's law of radiation. (You may recall from Chapter 6 that it was this law that opened the Pandora's box of quantum theory.) This law is a universal law of radiation for all black bodies. It allows the calculation of the thermal radiation of any black body object. The only necessary parameter for this equation is the temperature of the object. Using the same formula,

Figure 15.8 Thermal radiation spectrum of black body at 5780°K, which approximates the solar spectrum.

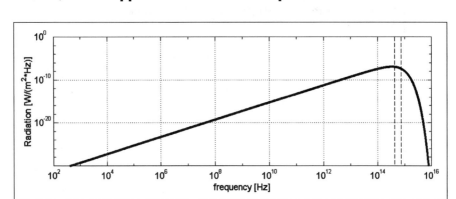

the spectrum of radiation from the sun can be calculated. Approximating the sun as a black body with a temperature of 5780°K, Figure 15.8 was created. The peak of radiation falls at about 350 THz, just below the visible red. (The human eye can see from about 430 THz to 750 THz.)

The temperature of the universe has been measured using this law. The temperature of the universe is due to the initial explosion commonly referred to as the big bang. This temperature is very low—about 3 degrees above absolute zero.

Radiated Thermal Noise

The existence of solar thermal radiation is great for life on Earth. However, thermal radiation of solar, earthly, and extraterrestrial origin also appears as noise to receiving antennas. In fact, the 3°K temperature of deep space was discovered to some extent by accident. Two scientists, A.A. Penzias and R.W. Wilson, at AT&T Bell Laboratories, were trying to determine the source of noise that was being collected by their microwave antenna dish. After eliminating all other possibilities, these scientists made the bold hypothesis that this noise was the thermal radiation of deep space. For this discovery, they received the Nobel Prize in physics.

Astronomers have measured the various sources of sky noise across the spectrum. Other sources of noise that interfere with antenna reception are solar radiation and galactic radiation. The atmosphere itself is also a source of noise, as is thermal radiation from the Earth. Figure

15.9 illustrates the many sources of electromagnetic noise present on the Earth.

CIRCUIT NOISE

There are several types of circuit noise, but I will discuss the most common type—that caused by the thermal movement of electrons in wires and resistors. This type of noise is referred to as *Johnson noise*, named after the researcher J.B. Johnson, who performed the first measurements of resistor noise in 1928. The random motion of electrons causes small random currents that appear as noise in circuits. Electrical current is proportional to the velocity of the electrons, not the acceleration. If you churn through the mathematics, you will find that the noise current is proportional to the square-root of the radiated noise, divided by the frequency. Hence, the spectral properties of the circuit noise are different from thermally radiated noise. Johnson noise is approximately constant across the frequencies used for electronics, and is therefore called white noise. The term white noise is an analogy to white light, which is an even mixture of all the colors of the visible spectrum. The noise voltage of a resistor is simply the noise current multiplied by the resistance.

CONVENTIONAL AND MICROWAVE OVENS

Microwave ovens work by bathing food in high-intensity radiation at 2.45 GHz. There is a popular myth that explains microwave ovens as operating at a special resonance of water molecules. In reality, this myth is just that, a myth. Referring to the Figure 15.2, you can see that there is no resonance of water at this frequency. The first resonant peak occurs above 1 THz, and the highest loss occurs well into the infrared. There is no special significance of 2.45 GHz, except that it is allocated by the FCC as being allowable for microwave oven usage. As is commonly known, microwave ovens differ from conventional ovens in that they heat the entire volume of a piece of food, whereas conventional ovens heat from the surface inward. It is this volume heating effect that allows microwave ovens to cook food very quickly. The reason microwave ovens perform as they do is readily seen by examining the frequency behavior of water, as shown in Figure 15.2. In the low GHz range, the skin depth or penetration depth of water is about 0.5 cm to 4.7 cm, depending on the salt content. Thus, microwaves will heat water to a

344 FREQUENCY DEPENDENCE

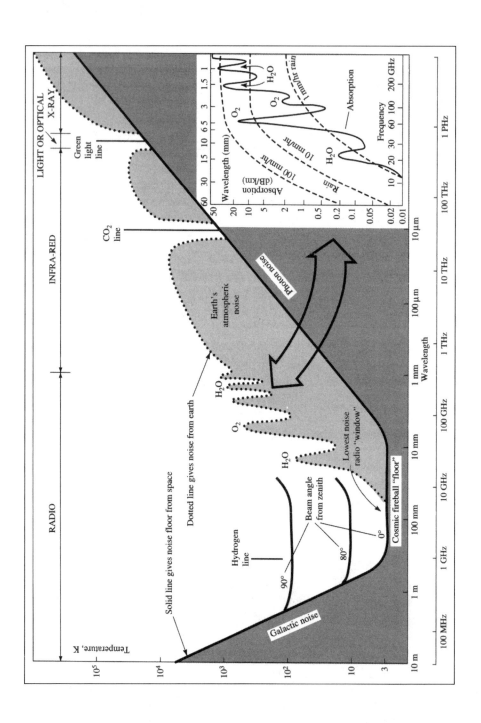

Figure 15.9 Sky noise from radio to X-rays. Inset illustrates attenuation by the atmosphere and by rain. Reproduced from Kraus, J.D., and D.A. Fleisch, *Electromagnetics with Applications*, 5th Edition, Boston: McGraw-Hill, 1999, p. 336, with permission of The McGraw-Hill Companies.

Figure 15.10 Voltage noise spectrum of a 1 kohm resistor at 37°C (100°F). The noise is white up into the infrared.

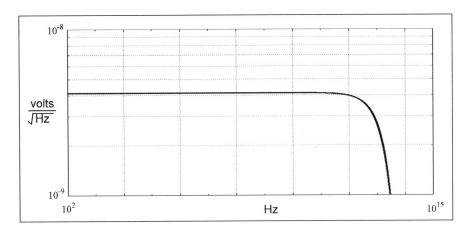

depth of well over an inch at these frequencies. You could even cook food at lower, RF, frequencies such as 100 MHz. There are two disadvantages to using lower frequencies. First, the oven would need to be 20 times larger so that a resonant cavity could be built. (For a radiation oven to work, the dimensions of the oven must be at least half of a wavelength.) Second, water is less lossy at 100 MHz, requiring longer heating times. On the other hand, you could use higher frequencies, say 100 GHz. The oven size could be made even smaller, but the problem with using this frequency is that the skin depth is too small. At 100 GHz the skin depth of water is about 1/100 to 1/64 inches. This type of microwave oven will produce surface heating only. The U.S. military is actually experimenting with radiation weapons at around 100 GHz. This radiation causes an immediate painful burn, but does not penetrate deeply into the body, so no long term harm is caused. From this analysis, you can see that the range of 1 GHz to 5 GHz is the optimal region for microwave oven–style heating. However, there is no special significance to 2.45 GHz. In fact, the unlicensed wireless data protocol called "Bluetooth" uses the same frequency of 2.45 GHz!

Conventional ovens cook food via infrared radiation from the heater coils, and by convection and conduction of the air inside the oven. Assuming that the coil is at a temperature of about 700°C, the radiation peak occurs at 40 THz. The skin depth of water at this frequency is about 30 µm, resulting in only surface heating. The heating resulting from contact with the hot oven air is also a surface effect.

BIBLIOGRAPHY AND SUGGESTIONS FOR FURTHER READING

Abrikosov, A. A., *Fundamentals of the Theory of Metals*, Amsterdam: North-Holland, 1988.

Agilent Technologies, *Making Precompliance Conducted and Radiated Emissions Measurements with EMC Analyzers*, Agilent Technologies, AN 1328.

Boithias, L., *Radio Wave Propagation*, Boston: McGraw-Hill, 1987.

DeWitt, D. P., and G. D. Nutter, *Theory and Practice of Radiation Thermometry*, New York: John Wiley, 1988.

Encyclopedia Britannica Inc., "Sunlight"; "Glass"; "Electromagnetic Radiation," *Encyclopedia Britannica*, Chicago: Encyclopedia Britannica Inc., 1999.

Fenske, K., and D. Misra, "Dielectric Materials at Microwave Frequencies," *Applied Microwave & Wireless*.

Henning, T., and H. Mutschke, "Low-temperature Infrared Properties of Cosmic Dust Analogues," *Astronomy and Astrophysics*, vol 327, pp. 743–754.

Hewitt, H., and A. S. Vause, *Lamps and Lighting*, New York: American Elsevier Publishing Company, Inc., 1966.

Hummel, R. E., *Electronic Properties of Materials—An Introduction for Engineers*, Berlin: Springer-Verlag, 1985.

Hutchinson, C., J. Kleinman, and D. R. Straw, Editors, *The ARRL Handbook for Radio Amateurs*, 78th edition, Newington, Conn.: American Radio Relay League, 2001.

Jackson, J. D., *Classical Electrodynamics*, 2nd Edition, New York: John Wiley & Sons, 1975.

Jonscher, A. K., *Dielectric Relaxation in Solids*, London: Chelsea Dielectrics Press, 1983.

Kittel, C., *Thermal Physics*, 2nd Edition, San Francisco: W. H. Freeman and Company, 1980.

Kittel, C., *Introduction to Solid State Physics*, 7th Edition, New York: John Wiley, 1996.

Kraus, J. D., and D. A. Fleisch, *Electromagnetics with Applications*, 5th Edition, Boston: McGraw-Hill, 1999.

Lide, D. R., Editor-in-chief, *CRC Handbook of Physics and Chemistry*, 72nd Edition, Boca Raton, Florida: CRC Press, 1991.

Littleton, J. T., *The Electrical Properties of Glass*, New York: John Wiley & Sons, 1933.

Murdoch, J., *Illumination Engineering from Edison's Lamp to the Laser*, New York: Macmillan Publishing Company, 1985.

Ordal, M. A., L. L. Lone, R. J. Bell, et al., "Optical Properties of the Metals Al, Co, Cu, Au, Fe, Pb, Ni, Pd, Pt, Ag, Ti, and W in the Infrared and Far Infrared," *Applied Optics*, Vol. 22, No. 7, April 1, 1983.

Page, L., and D. Wilkinson, "Cosmic Microwave Background Radiation," *Reviews of Modern Physics*, Vol. 71, No. 2, 1999.

Palik, E. D., Editor, *Handbook of Optical Constants of Solids*, Orlando, Florida: Academic Press, 1985.

Papoulis, A., *Probability, Random Variables and Stochastic Processes*, New York: McGraw-Hill, 1991.

Penzias, A. A., *Measurement of Cosmic Microwave Background Radiation*, IEEE Transactions on Microwave Theory and Techniques, Vol. MTT-16, NO.9, September 1968.

Pozar, D. M., *Microwave Engineering*, 2nd Edition, New York: John Wiley, 1998.

Ramo, S., J. R. Whinnery, and T. Van Duzer, *Fields and Waves in Communication Electronics*, 2nd Edition, New York: John Wiley, 1989.

Ray, P., "Broadband Complex Refractive Indices of Ice and Water," *Applied Optics*, vol. 11, p. 1836–1844, 1972.

Roussy, G., and J. A. Pearce, *Foundations and Industrial Applications of Microwave and Radio Frequency Fields*, England: John Wiley & Sons Ltd., 1995.

Segelstein D. J., "The complex refractive index of water," M.S. Thesis, University of Missouri-Kansas City, (1981).

Seraphin, B. O., *Optical Properties of Solids: New Developments*, Amsterdam: North-Holland Publishing Company, Chapter 12, 1976.

Siegel, R., and J. R. Howell, *Thermal Radiation Heat Transfer*, 2nd Edition, New York: McGraw-Hill, 1981.

Solymar, L., and D. Walsh, *Lectures on the Electrical Properties of Materials*, 5th Edition, Oxford: Oxford University Press, 1993.

Straw, R. D., Editor, *The ARRL Antenna Book*, 19th Edition, Newington, Conn,: American Radio Relay League, 2000.

Thuery, J. *Microwaves: Industrial, Scientific, and Medical Applications*, Boston: Artech House, 1992.

Web Resources

http://omlc.ogi.edu/spectra/index.html
http://urania.astro.spbu.ru/JPDOC/1-entry.html
http://www.crystran.co.uk/
http://www.chipcenter.com/circuitcellar/august99/c89r4.htm

Appendix A ELECTRICAL ENGINEERING BOOK RECOMMENDATIONS

Computer Networks

Halsall, F., *Data Communications, Computer Networks and Open Systems*, 4th Edition, Reading, Mass.: Addison-Wesley, 1996.

Halsall, F., *Multimedia Communications: Applications, Networks, Protocols, and Standards*, Reading, Mass.: Addison-Wesley, 2000.

Tanenbaum, S., *Computer Networks*, 3rd Edition, Upper Saddle River, NJ: Prentice Hall, 1996.

Electrical Engineering

Cogdell, J. R., *Foundations of Electrical Engineering*, 2nd Edition, Upper Saddle River, NJ: Prentice-Hall, 1995.

Practical Electronics

Horowitz, P., and W. Hill, *The Art of Electronics*, Cambridge, Mass.: Cambridge University Press, 1980.

Designing with Transistors

Sedra, A. S., and K. C. Smith, *Microelectronics Circuits*, 2nd Edition, Fort Worth, Texas: Holt, Rinehart, and Winston, 1987.

Practical Tips for Analog Design

Williams, J., Editor, *Analog Circuit Design: Art, Science, and Personalities*, Newton, Mass.: Butterworth–Heinemann, 1991.

Pease, R. A., *Troubleshooting Analog Circuits*, Newton, Mass.: Butterworth–Heinemann, 1991.

Practical Electrical Engineering

Williams, T., *The Circuit Designer's Companion*, Newton, Mass.: Butterworth–Heinemann, 1991.

Digital Circuit Design

Haznedar, H., *Digital Microelectronics*, Redwood City: Benjamin/Cummings Publishing Co., 1991.

High-Speed Digital Design

Johnson, H., and M. Graham, *High-Speed Digital Design: A Handbook of Black Magic*, Englewood Cliffs: Prentice-Hall, 1993.

High-Speed Digital PCB Layout

Montrose, M. I., *Printed Circuit Board Design Techniques for EMC Compliance—A Handbook for Designers*, 2nd Edition, New York: IEEE Press, 2000.

Practical EMC

Montrose, M. I., *EMC and the Printed Circuit Board—Design, Theory, and Layout Made Simple*, New York: IEEE Press, 1999.

Williams, T., *EMC for Product Designers*, Oxford: Butterworth–Heinemann Ltd, 1992.

EMC Theory

Paul, C. R., *Introduction to Electromagnetic Compatibility*, New York: John Wiley & Sons, 1992.

Shielding and Cabling

Ott, H. W., *Noise Reduction Techniques in Electronic Systems*, 2nd Edition, New York: John Wiley, 1988.

RF Engineering

Shrader, R. L., *Electronic Communication*, 6th Edition, New York: McGraw-Hill, 1990.

Krauss, H. L., C. W. Bostian, and F. H. Raab, *Solid-State Radio Engineering*, New York: John Wiley, 1980.

Hagen, J. B., *Radio-Frequency Electronics: Circuits and Applications*, New York: Cambridge University Press, 1996.

Bowick, C., *RF Circuit Design*, Carmel, Indiana: SAMS Publishing, 1995.

Hutchinson, C., J. Kleinman, and D. R. Straw, Editors, *The ARRL Handbook for Radio Amateurs*, 78th Edition, Newington, Conn.: American Radio Relay League, 2001.

Microwave Engineering

Pozar, D. M., *Microwave Engineering*, 2nd Edition, New York: John Wiley, 1998.

Matthaei, G., E. M. T. Jones, and L. Young, *Microwave Filters, Impedance-Matching Networks, and Coupling Structures*, Boston: Artech House, 1980.

Practical Antennas

Straw, R. D., Editor, *The ARRL Antenna Book*, 19th Edition, Newington, Conn.: American Radio Relay League, 2000.

Carr, J. J., *Practical Antenna Handbook,* Fourth Edition, New York: McGraw-Hill, 2001.

Antenna Theory

Kraus, J. D., *Antennas*, 2nd Edition, Boston: McGraw-Hill, 1988.

King, R. W. P., and C. W., Harrison, *Antennas and Waves: A Modern Approach*, Cambridge, Mass.: The M.I.T. Press, 1969.

Electromagnetics (in Order of Difficulty)

Kraus, J. D., and D. A. Fleisch, *Electromagnetics with Applications*, 5th Edition, Boston: McGraw-Hill, 1999.

Ramo, S., J. R. Whinnery, and T. Van Duzer, *Fields and Waves in Communication Electronics*, 2nd Edition, New York: John Wiley, 1989.

Heald, M., and J. Marion, *Classical Electromagnetic Radiation*, 3rd Edition, Fort Worth, Texas: Saunders College Publishing, 1980.

Griffiths, D. J., *Introduction to Electrodynamics*, 3rd Edition, Upper Saddle River, NJ: Prentice Hall, 1999.

Vanderlinde, J., *Classical Electromagnetic Theory,* New York: John Wiley & Sons, 1993.

Jackson, J. D., *Classical Electrodynamics,* 2nd Edition, New York: John Wiley & Sons, 1975.

Physics

Halliday, D., R. Resnick, and J. Walker, *Fundamentals of Physics*, 6th Edition, New York: John Wiley & Sons, 2000.

Feynman, R. P., R. B. Leighton, and M. Sands, *The Feynman Lectures on Physics Vol I: Mainly Mechanics, Radiation, and Heat*, Reading, Mass.: Addison-Wesley Publishing, 1963.

Feynman, R. P., R. B. Leighton, and M. Sands, *The Feynman Lectures on Physics Vol II: Mainly Electromagnetism and Matter*, Reading, Mass.: Addison-Wesley Publishing, 1964.

Epstein, L. C., *Thinking Physics—Is Gedanken Physics; Practical Lessons in Critical Thinking*, 2nd Edition, San Francisco: Insight Press, 1989.

INDEX

λ/20 rule, 287
3W rule, 286
4-vectors, 123–124
5/5 rule, 269–270

Aether, 27, 86
Alternating current
　power radiated from, 102–103
　slow, 102–103
Ampere's law, 84
Analog signals, 12
Angular resolution, 315–316
Antenna arrays, 316–319
Antenna pattern, 236
Antennas. *See also* Dipole antennas; Loop antennas; Monopole antennas
　AM radio, 249
　axis of symmetry of, 237
　car radio, 244
　directed, 246, 247
　effective cross-section, 245–246
　FM radio, 240–241
　input impedance, 232–236
　loop stick, 249
　passive elements, 246, 247
　polarization, 239–240
　radiation patterns versus length, 236–239
　reflecting dishes, 307–311
　resistive losses in, 244–245
　slot, 187
　"smart," 318–319
　traveling wave, 246–247
　wire radius effects, 235, 244–245
Antiparticles, 131
Antiresonance, 274
Asymmetric strip-line, 286
Atoms, 30
Audio speakers, 67–68
Aurora Borealis, 63

Babinet's principle, 324–325
Back-emf, 75
Balanced signaling, 302–303
Battery exciting transmission line, 155–157, 215–217
Beamforming, 318
Bethe, Hans, 325
Beverage antenna, 212–213
Bio-Savart law, 52
Biological sensitivities, 41
Black body radiation, 341–342
Bonds, atomic, 30
Born, Max, 128–129
Break frequency, 298–299

Cables
　above break frequency, 299–302
　asymmetric currents on, 295
　below break frequency, 297–298
　break frequency of, 298–299
　coaxial, 175, 297, 299–300
　electrical configurations for, 302–303
　radiation from, 290–296
　ribbon, 300
　shields on, 296–297, 298, 300
　triaxial, 298
　twisted pair, 301–302
Capacitance, 39, 139
Capacitive coupling, 257
Capacitors
　AC voltage across, 85
　bulk, 279
　characteristics per type, 143
　defined, 36
　electric fields surrounding, 44, 45
　energy storage in, 93
　hidden schematic of, 142
　impedance of, 79
　low inductance arrays, 277
　microstrip, 220

353

Capacitors (cont.)
　organic semiconductor electrolytic, 276
　reactive impedance of, 139, 142
　surface mount, 277
Chamfered corners, 282–283
Charged particles, 131
Charges
　accelerating, 96–99, 126
　biological sensitivities to, 41
　constant velocity, 96–97, 99, 101
　force fields around, 26
　half-life of, 34
　magnetic field generated by, 51–52
　oscillating, 99–100
　polarities of, 25
　polarization of, 334–335
　static point, 94–95
　stationary, 124–125
　"theft" of, 33
　types of, 30
Cladding, 205
Clock signal routing, 284, 285
Coaxial cables, 175, 297, 299–300
Coherent light, 20, 205
Collapse of the wave function, 128
Common impedance coupling, 267–268
Common mode chokes, 294–296
Common mode radiation, 290–294
Components. See also Capacitors; Inductors; Resistors
　measurements on, 150
　selection considerations, 151–152
Conducted interference, 256–257
Conductivity of materials, 47
Conductor bundling, 209
Constitutive relations, 87
Copper
　conductivity of, 47
　electric properties by frequency, 333
　low-frequency permeability of, 73
　radiation incident on, 186, 191
　reflection coefficient of, 184
　relaxation constant, 33–34
　shielding effectiveness of, 187, 192
　skin depth of, 185
　wave impedance in, 183
Copper wire, 32
Coriolis force, 59–63
Corona, 43
Coulomb's law, 25, 26
Cross-coupling, 252–253
Cross product right hand rule, 58
Crosstalk, 257–259, 286

Current. See Alternating current; Direct current
Cutoff frequency, 196

D (electric displacement field), 87
Damping frequency, 332
DC motors, 58–59
de Broglie, Louis, 128
Degaussing, 71
Delay, 8–9
Diamagnetism, 69–70, 72
Dielectric constant, 39
Dielectrics, 18, 38–39, 334–335
Differential mode radiation, 290–291
Diffraction
　defined, 321
　Fraunhofer, 325
　Fresnel, 325–326
　radio wave, 326–327
Digital signals, 12
Diodes, 28–29
Dipole antennas
　basics of, 93–94, 229–230
　effects of ground on, 241
　feed reactance of, 242
　folded, 248–249
　input impedance of, 232–236
　near-field/far-field boundary, 105
　radiating field, 105
　radiation patterns
　　versus height, 243
　　versus length, 236–239
　radiation resistance of, 231–232, 233, 242
　reactive field, 105
　wave impedance, 106
Dipoles
　electric, 229–230
　magnetic, 56–57, 230–231
　static charge, 95
Dirac, P.A.M., 131
Direct current, 31–32, 63–64
Directional couplers, 222–225
Displacement current, 36, 85
Distributed circuit techniques, 218, 219
Distributed grounds, 269–270
Diversion shielding techniques, 266–267
Doppler effect, 115
Drift velocity, 32
Dual strip-line, 285–286

Eddy currents, 83–84
Electret, 71

INDEX

Electric displacement field (D), 87
Electric fields
  around capacitors, 44, 45
  around wire
    alternating current, 102–103
    direct current, 46
  defined, 25
  illustration conventions, 25–26
  magnitude of, 26
  static, 94–95
Electric guitars, 80
Electric shocks, 41
Electrical length, 8, 14, 91–92
Electromagnetic compatibility (EMC)
  defined, 2, 89, 251
  obstacles to, 252–259
  problem categories, 2, 255–257
Electromagnetic devices, 67–68, 79–81
Electromagnetic fields
  energy in, 130–131
  field power density of, 105
  types of, 89–91
  zones of, 107–108
Electromagnetic radiation
  cause of, 99, 113
  reflection of, 182–190
  spectrum of, 5
Electromagnetic waves
  basic component of, 1
  characteristics of, 6–7
  propagation predicted, 85
  speed of, 209–210
Electromagnetics, 3
Electronics, 9, 16–17
Electrons
  defined, 9, 55
  free space, 131
  kinetic energy of, 29
  mobility in metals, 33
  orbiting, 69–70
  polarity of, 30
  quantum electrodynamic, 132
  spin of, 54, 55, 70
Electrostatic induction, 34–38
End effects, 232
Equations of state, 87
Evanescent waves, 203

Far fields. *See* Radiating fields
Far infrared, 16
Faraday cages, 188
Faraday's law, 76
Ferrimagnetic materials, 72

Ferrites, 72, 83–84
Ferroelectric effect, 71
Ferromagnetic shielding, 298
Ferromagnetism, 71, 72
Feynman, Richard, 127
Fiber optics, 20–21, 204–205
Field detector, 37–38
Field power density formula, 105
Fluorescent lights, 43
Focal point, 312
Force fields, 25–26
Fraunhofer zone, 107–108
Free radicals, 19
Frequency-to-wavelength formula, 4
Fresnel zone, 107–108
Frustrated waves, 203–204

Geometrical optics approximation, 307
Glass, 47, 335–337
Glow tube meter, 38
Gravity, 26–27, 113
Ground bounce, 267–269
Ground (earth), 338, 339, 340
Grounds
  configurations for, 269–271
  ground loops, 287–290
Group velocity, 196–197
Guard traces, 266

H-field, 87
Heisenberg Uncertainty Principle, 129–130
Hertz, Heinrich, 85
"Hidden schematic," 14, 139
High-frequency regime, 14
Holes, 33
Huygens-Fresnel wave propagation principle, 323
Hybrid grounds, 270–271
Hysteresis, 82

Ideal Gas law, 87–88
Image planes
  optical, 313
  printed-circuit, 285
Imaging, 313–315
Imaging devices, 315–316
Impedance
  50 ohm standard, 174–175
  defined, 78
  imaginary, 78
  intrinsic material, 182
  parasitic, 139

356 INDEX

Index of refraction, 205
Induced interference, 255–258
Inductance, 76–77, 139
Inductive coupling, 258
Inductors
  AC across, 78
  cores for, 81–84
  defined, 76
  discharging, 77–78
  energy storage in, 78, 92–93
  frequency responses of, 144–145, 147–148
  hidden schematic of, 146
  impedance of, 79, 139
  microstrip, 220
  parasitic capacitance of, 78, 143
Inertial frames of reference, 113
Infrared effects, 16–18
Insulators, 18, 38–39, 334–335
Intrinsic material impedance, 182
Ionosphere, 201, 334
Ions, 30

Johnson noise, 343

Kirchhoff's laws, 10–11

Lamps, incandescent, 206
Lasers, 20, 205–206
Layout diagrams, 259
Lenses, 311–313
Lenz's law, 75–76
Light
  dual nature of, 128–129, 321
  Huygens-Fresnel propagation principle, 323
  speed of, 9, 114–115
Lightning, 42–43
Lightning rods, 44
Litz wire, 209
Lobes, 237
Local decoupling, 281–282
Longitudinal balance, 302
Loop antennas
  basics of, 93–94, 230–231
  characteristics per size, 236
  multiturn, 249
  radiation resistance, 232, 234
Lorentz-FitzGerald contraction, 118–121
Lorentz force law, 62
Lorentz transformation, 117
Loudspeakers, 67–68
Lumped element design, 10, 12, 217–218

Magnetic dipoles, 53–59
Magnetic fields
  around magnets, 56
  around moving electrons, 52
  around wires, 51–52
  Earth's, 54–55, 57
  general law governing, 66
  nature of, 51
  vector potential of, 68–69, 133
Magnetic materials classified, 72
Magnetic monopoles, 51
Magnetic permeability of materials, 73
Magnetic vector potential, 68–69, 133
Magnets, 71
"Mains," 37
Marconi, Guglielmo, 86
Matter, 30, 328
Maximum power transfer law, 173
Maxwell's equations, 85–86, 111–112, 125
Metals, 30–31, 33, 332–334
Metric equation, 120
Microstrip transmission lines
  characteristic impedance of, 172, 261
  common forms of, 260
  distributed circuit techniques with, 218–219
  frequency limitations of, 200
  grounding example, 293
  low-pass filter, 17
  paths of least impedance in, 262–264
  wave velocity of, 261–262
Microwave effects, 16
Microwave frequency range, 218
Minkowski, Hermann, 120
Mirrors, 181
Molecules, 30
Momentum-energy vector, 124
Monopole antennas
  1/4 wave, 241–243
  5/8 wave, 243–244
Monopoles, 230
Motion, 113–114
Motors, 58–59

Near fields. *See* Storage fields
Neutrons, 30
Newton's laws, 113
Nodes, 10–11
Noise, 20, 31, 342–344
Northern Lights, 63
Norton wave, 212

INDEX

Ohm's law, 9
One-port devices, 221
Optical theories, 18–19
Orbitals, 30
Ovens, 343–346
Overshoot/undershoot, 171

Paramagnetism, 70–71, 72
Parasitic capacitance, 139
Parasitic inductance, 139
PCB stackup, 286
Period formula, 4
Permeability, 73
Permittivity of materials, 47
Photon energy formula, 4
Photonics, 20–21
Photons, 1, 18–19, 127
Planck, Max, 126–127
Planck's constant, 127
Plasma frequency, 334
Positrons, 131
Potential energy, 29
Power, AC, 102–103
Power lines, 174, 209
Power supply decoupling
    capacitor placement, 278
    capacitors for, 273–278
    challenges in, 276
    designing, 278–282
    goals of, 271–273
    high frequency, 281
    local, 281–282
    low frequency, 278
    low-pass filter for, 272, 273
    modeling of, 274
    need for, 271
    power supply, 281
    two-layer board, 290
Printed circuit boards, 259–260
Printed circuits
    chamfered corners, 282–283
    clock signal routing, 284, 285
    cross-coupling within, 252–253
    ground decoupling, 292
    ground loops, 287–290
    grounding configurations, 269–271
    guard traces, 266
    IC decoupling, 292
    image planes for, 285
    layout examples, 287, 288, 289, 290, 291–292
    layout rules, 264–265
    layout tips, 286–287

power supply decoupling, 271–282, 290, 292
    power/ground planes on, 277
    routing guidelines, 285–286
    shielding on, 265–267
    transmission line, 172
    transmission line terminations, 283–284
Propagation modes, 197–199
Proper length, 115
Proper mass, 124
Proper time, 115
Proper velocity, 122–123
Protons, 30

Quantum electrodynamics, 127–134
Quantum physics, 126–129, 134
Quantum strangeness, 127–128

Radiated interference, 255
Radiating fields
    AC source, 105–107
    accelerated charge, 97–108
    characteristics summarized, 107
    circuits generating, 93–94
    defined, 89–91
Radiation. *See* Electromagnetic radiation
Radiation resistance, 231–234
Ray theory, 18
Reactance, 78
Reactive fields. *See* Storage fields
Reciprocity law, 231
Reflecting dish antennas, 307–311
Reflection coefficient formula, 160
Reflections in RF systems, 172–173
Relativistic mass, 124
Relativity theory, 111–125, 134
Relaxation frequency, 332
Relaxation time, 33
Relays, 79
Renormalization, 132
Resistance, 32
Resistors
    formula response of, 140–142
    hidden schematic of, 140
    pull-up, 171–172
    types of, 139–140
Resonant cavities, 204
RF chokes, 151, 273
RF coupling capacitors, 150
RF decoupling, 282
Ribbon cables, 300

Right hand rule, 53
Ringing, 252

S-parameters, 221–223
Saturation, 82
Schematic diagrams, 259
Schroedinger, Erwin, 128–129
Schumann resonance, 204
Self-compatibility, 251
Semiconductors, 33
Shelves, 171
Shielding effectiveness, 185–186, 187
Shields
    basics of, 181
    electrical thickness of, 192, 194
    far field, 184–190
    Faraday cages, 188
    gaskets, 189–190
    grounding of, 190–191
    holes in, 186–188
    honeycomb, 189
    mesh, 188, 189
    microwave oven, 188–189
    near field, 190
    tubular, 189
Signal integrity, 251–252
Signals, digital
    anomalies, 253
    frequency spectra of, 254
    rise times, 252
Simultaneity, 117–120
Simultaneous switching noise, 269
Single-ended cabling, 302
Skin depth, 183, 185, 299–300
Skin effect, 183–184, 332
Skin resistance, 41
Skip-mode, 201–202
Sky noise, 344
Solenoids, 54
Sommerfeld-Goubau wave, 214
Space-time, 115–124
Space travel, 121–123
Spectrum
    electromagnetic, 5
    visible, 319
Spin, 54, 55, 70, 131
Square waves, 12
St. Elmo's fire, 44
Standing waves, 177–178
Static electricity, 39–42
Storage fields
    AC source, 105–107
    characteristics, 107

circuits generating, 93
defined, 89–91
Stray fields, 253
Stretch factor, 116–117
Strip-line, 286
Superconductors, 133
Superparamagnetic materials, 72
Surface acoustic wave devices, 210
Surface mount devices, 218
Surface waves
    coupled, 214–217
    defined, 210
    propagation of, 211–217

Telephone line directional couplers, 224–225
Telephone ringers, 79–80
TEM (transverse electric and magnetic) mode, 200, 214
Temperature, 31, 339–340
Tensor, 124
Tesla, Nikola, 86
Thermal radiation, 126–127, 338–342
Thompson, G.P., 128
Time desynchronization, 117–120
Time dilation, 117–118
Time domain reflectometers, 175–176
Time domain sampling, 317–318
Time retardation, 96
Time shielding techniques, 265–266
Tolerance, 139
Torque, 57–58
Transfer impedance, 299–300
Transformers, 81, 83
Transmission coefficient formula, 160
Transmission line effects, 12, 14–16
Transmission lines. *See also* Microstrip transmission lines
    analog signals on, 165–167
    battery exciting, 155–157, 215–217
    characteristic impedances of, 155–157, 175–177
    current loops in, 294
    defined, 153
    digital signals on, 163–164
    eighth-wavelength, 218–220
    high-power, 174
    impedance matching
        digital systems, 171–172
        RF systems, 172–174
    impedance transformations by, 167–171
    length effects of, 167–171
    losses in, 173

INDEX 359

lossless, 154
maximum power transfer law, 173
measuring impedances of, 175–177
models of, 154–155
near field surrounding, 258
printed circuit board, 172
quarter-wavelength, 168–169
reflections in, 159–167, 172–173
sinusoidal signals on, 178
standing waves on, 177–178
strip-line, 172, 200
surface waves on, 214–217
traditional, 214
waveguide model of, 157–159
waveguide modes on, 199–201, 217
waveguides versus, 217
Traveling wave antennas, 246–247
Triaxial cables, 298
Triboelectric effect, 41, 45
Tunneling, 17–18, 204
Twin paradox, 123
Twisted pair cables, 301–302
Two-port devices, 221–223

Ultraviolet rays, 19

Vector network analyzer, 222–224
Virtual electron/positron pairs, 131–132
Virtual photons, 130–131
Vision, 319
Voltage, 28, 29
Voltage nodes, 10–11

Water
  absorption coefficient, 18
  charged molecules, 40
  electrical properties of, 47, 335–336
  low-frequency permeability of, 73
  molecular asymmetry, 39
  skin depth of, 343, 345
Wave antenna, 212–213
Waveguides
  antennas, 202–203
  cutoff frequency, 196
  cylindrical, 17
  defined, 194–195
  Earth's, 201–202
  evanescent waves in, 203
  fiber optics as, 204–205
  grounded coplanar, 266, 267
  minimum size, 194
  propagation modes within, 197–199
  transmission lines versus, 217
  tunneling in, 204
Wavelength division multiplexing, 205
Wavelength-to-frequency formula, 4
White noise, 20, 31
Wires
  as conductors, 10
  current-carrying, 65–67
  heat losses from, 101
  hidden schematic of, 149

X-rays, 19–20, 98

Yagi-Uda array, 246, 247

Zenneck wave, 212